Coral Disease Identification and Monitoring Manual

Student Study Book and Manual

This manual has been published by Conservation Diver Foundation.

©2022 Conservation Diver, all rights reserved

Written by Elouise S Haskin

With Contributions by Chad Scott and Spencer Arnold

All pictures, graphics, and text contained herein remain the sole copyright property of the author and may not be duplicated or distributed without explicit prior written permission from the author.

This book may not be reproduced, copied, or replicated without explicit written permission from the creator or Conservation Divers Foundation.

This manual only provides a background and a broad outline for conducting the Conservation Coral Disease Certification Course, this does not serve as an alternative to learning the course with an instructor and completing the underwater skills development work. Anybody wishing to conduct the EMP should do so only under the close supervision of a trained professional to prevent harm to the environment or person. The authors of this guide and the Conservation Divers Foundation are not responsible for any injuries or problems sustained while participating in the Coral Disease Course. SCUBA diving is an inherently dangerous sport and is done at your own risk.

Please cite this document as:

Haskin, E., S. (2022) **Coral Disease Identification and Monitoring Manual.** *Conservation Diver Foundation,* Colorado, USA. 89 pp
published version ISBN: 978-1-7326925-2-7
Ebook ISBN: ISBN: 978-1-7326925-3-4

Contents

Acknowledgements 6
Coral Disease Identification and Monitoring Description 6
Prerequisites 7
Course Outline 7

Chapter 1: Introduction to Coral Health, Disease and Surveys 9
1.1. Introduction to Coral Diseases 9
1.2. How Diseases Become Problematic 10
1.3. Coral Disease Terminology Basics: 12

Chapter 2: Predation 19
2.1. Crown of Thorns Sea Star (COTS) 19
2.2. Corallophila Snails 24
2.3. Wentletrap Snails 24
2.4. Drupella Snails 25
2.5. Damselfish 27
2.6. Butterflyfish - 28
2.7. Triggerfish 30
2.8. Parrotfish 31

Chapter 3: Competition 33
3.1. Macroalgae/Nutrient Indicator Algae 33
3.2. Cyanobacteria/Filamentous Mats 34
3.3. Tunicates/Ascidian Overgrowth 36
3.4. Sponge Overgrowth 38
3.5. Competition between Coral Colonies 39

Chapter 4: Other Symptoms 41
4.1. Unexplained Bleaching 41
4.2. Pigmentation Response 42
4.3. Growth Anomalies 43
4.4. Sedimentation 44
4.5. Trematodiasis - 46
4.6. Physical Disturbance 47
4.7. Trash/Rubbish 47

Chapter 5: Diseases of the Indo-Pacific 49
5.1. White Syndrome 49
5.2. Black Band Disease 51
5.3. Brown Band Disease 53
5.4. Skeletal Eroding Band 55
5.5. Pacific Yellow Band Disease 57
5.6. Atramentous Necrosis 58

Chapter 6: Diseases of the Wider Caribbean Region 61
6.1. Aspergillosis 61
6.2. Black Band Disease 62
6.3. Caribbean Ciliate Infection 62
6.4. Caribbean Yellow Band Disease/Yellow Blotch Disease 64
6.5. Dark Spots Disease... 65
6.6. White Band Disease Type I and Type II 66
6.7. White Patch Disease 68
6.8. White Plague Disease 68
6.9. Stony Coral Tissue Loss Disease... 69

Chapter 7: Data Collection 71
7.1. Belt-Transect 71
7.2. Survey Slates 73
7.3. Quadrat Survey 75
7.4. Tracking Colonies 77
7.5. Data use 78

Chapter 8: Coding Accurately 79
Appendix A..82
Appendix B..83
Glossary 85
References..86

Acknowledgements

This manual would not have been possible without the help of many people. In particular the author would like to thank the Conservation Diver team, Chad Scott, Spencer Arnold, Rahul Mehrotra, Pau Urgell Plaza, George Bevan, Kirsty Magson and Leon Haines for proof reading, providing photos and making suggestions which helped build this learning resource.

Coral Disease Identification and Monitoring Description

This course is designed to give experienced divers a deeper understanding of coral reefs, and the threats they face from the many different diseases they've been observed to suffer from. The aim of this manual is to introduce the theory and practical considerations of surveying coral reef sites for disease. This course builds off the knowledge gained during the Conservation Diver Ecological Monitoring Program (EMP) certification, and will provide divers with the techniques needed to track reef health in more sophisticated detail than previously explored.

After completing this course, students will know how to perform the reef research methods of the Coral Disease Survey and will be able to collect data to a standard that will allow them to contribute to local and global databases. Students will learn how to identify additional threats to reef health, explain why they happen and what they represent. They will learn specifically about symptoms affecting the corals of the Indo-Pacific, enabling their identification and monitoring. Additionally, experience working with survey equipment underwater will be acquired, as well as further training in navigation skills and buoyancy. Students who complete the training can then volunteer to assist with the regular disease surveys, and utilize their knowledge in other parts of the Indo-Pacific.

Prerequisites

The following is a list of prerequisites for the completion of the Conservation Diver Coral Disease Identification and Monitoring course:
- Be an **Advanced level diver**, or show proficiency in good buoyancy (your Advanced Course dives must have included Peak Performance Buoyancy, Deep Dive, and Navigation).
- All students must **successfully demonstrate proficiency in buoyancy skills**. Including but not limited to:
  - Basic Hovering
  - Diving with fins above head level
  - Non-disturbance of substrate (silt, etc)
  - Fin Pivot
  - Ability to be stationary without anchoring
- Have been **certified in the Conservation Diver certificate: Ecological Monitoring Program (EMP).**
- Have been **certified in the Conservation Diver certificate: Coral Taxonomy and Identification.**
- Review all Safety Procedures and understand the particular risks of the areas you will be diving.
- Sign all health and liability waivers.

Course Outline

The following course is taught over the course of one day, assuming you are already trained and certified in Ecological Monitoring Program and Coral Taxonomy Identification courses. This single day will cover the theoretical knowledge during the lecture and practicing the necessary survey skills underwater during a survey dive. It is recommended that at least two practical dives are completed to ensure all course material is covered and questions can be addressed before data collections commence.

If the student has acquired the skills for the prerequisite topics, the course can be completed in one day, as shown in the table below, however if more practice is needed to complete the survey to a satisfactory level (data collected must be comparable with that of the instructor), more dives may be necessary.

	Description	Topics/skills covered
Day 1		
Lecture	1. "Coral Disease Identification and Monitoring"	- Basic coral threats - Introduction to coral lesions; borders, shapes and progression - Coral disease variation and identification - Underwater survey methods with transect lines - Data input and uses
Dive	1. Coral Disease Survey	- Navigating and reeling out survey line - Conduct disease survey - Identifying and describing compromised health states of corals

Chapter 1: Introduction to Coral Health, Disease and Surveys

1.1. Introduction to Coral Diseases

Currently, coral diseases are one of the leading causes of reef decline globally. Understanding the impact of coral disease is important for reef managers in order to better enable them to protect and restore local ecosystems and resources.

Coral disease research is still in its infancy, with the first record of disease in coral only reported in 1973. Since that time, the prevalence and mortality caused by coral diseases seems to be increasing in many locations around the globe, including those relatively isolated from centres of development. Outbreaks of disease have been most destructive in areas of the Caribbean, beginning with the outbreak of White Syndrome throughout the 1970's, and a more recent outbreak of Stony Coral Tissue Loss Disease that began in 2018. To date, around 95% of hard corals in the Caribbean have already been lost, most notably those from the fast growing and structurally complex Acroporiid generas. In the time of writing, outbreaks have been witnessed for the first time in the biogeographically remote Hawaiian Islands and other isolated areas of the Indo-Pacific.

A *Diploastrea heliopora* coral with Pacific Yellow Band Disease. Photograph: Elouise Haskin

1. Introduction to coral health, disease and surveys

Regular data collection is enormously valuable for reef management, and has already played major roles in developing management strategies across many environmental fields. This manual focuses on understanding the causes of coral disease, identifying threats that could lead to disease outbreaks, identifying diseases, and collecting field data or monitoring reefs for coral disease.

By the term 'coral disease', we are referring to any describable symptom of a coral that affects the corals ability to grow or reproduce. Technically speaking, according to this definition, everything discussed in this book falls under the category of a coral disease. However, for the sake of simplicity and easy understanding, we divided the categories into four groups: predation, competition, disease and 'other'.

1.2. How Diseases Become Problematic

Coral diseases play crucial roles in maintaining healthy coral populations by selectively removing weakened colonies and introducing evolutionary pressures for adaptation. Stressed corals can respond in a variety of ways to these stresses, including reduced fecundity during spawning events, the release of stress related amino acids or the expression of fluorescent proteins, or a coral disease may develop. These are some of the possible paths of a stressed coral, all of which contribute to the decline (directly or indirectly) of genetically weaker individuals from the gene pool, leaving the fittest to succeed and sexually reproduce. This succession of healthy animals in a population is often referred to as 'survival of the fittest'- the catch phrase that was coined in Charles Darwin's On The Origin of Species; a phrase that captures the fundamental idea of his theories on natural selection.

How is it that coral diseases have suddenly become more prominent and problematic? The answer may lie a little bit closer to home than we were all hoping. Within the last century our technology has revolutionised the way we live and our global population has exploded as a result. We eat more, do more, and pollute more than ever. Our impact on the planet has been so profound that many academics have come to use the term Anthropocene (which translates to the 'age of man') to describe the current epoch we live in; a period of earth's history where humans have been the defining factor in Earth's changing climate.

To name some examples:
- Increased sedimentation levels due to agricultural activities, deforestation, and land development
- Ocean acidification due to increased atmospheric carbon dioxide levels

> It is important to remember that coral disease research is still a relatively new area of study, and even amongst professionals, diseases can sometimes be near-impossible to identify in the field. Data collection is the name of the game, and contributions by coastal managers, researchers and citizen science projects can help make great leaps across gaps in our knowledge bases.

Introduction to coral health, disease and surveys

- Increased sea surface temperatures (SST) due to increased levels of greenhouse gas (GHG) emissions
- Increased nutrient levels due to improper agriculture and sewerage management
- Increased pollutants on reef ecosystems due to improper land management and waste runoff
- Marine debris in the form of plastics and other derivatives of the petro-chemical revolution.

Climate change is a reality, and at present is happening so quickly that many species are forced to adapt to these rapidly shifting conditions or face extinction. Coral reefs as fragile and, susceptible ecosystems are centre-stage when it comes to habitat degradation. So far, many have been unable to adapt to change, leading to large scale loss of corals and population shifts. The way these matters affect corals will be discussed in detail throughout this manual. As reef managers and ecologists, it is important to be able to identify and grasp these problems on a local basis in their infancy, to ensure appropriate monitoring and management entails.

Some coral diseases have geographical limitations, yet confusion still occurs amongst these different regions. For example, there are many 'white diseases'. White Band Disease (WBD) is present in the Caribbean, affects species within the Acropora genus and has so far has not been recorded elsewhere. The Indo-Pacific however, contains similar diseases referred to as White Syndrome (Formally White Band, White Pox, etc.) that affect Acropora amongst many other genera.

This topic at first, and understandably, can be quite perplexing. However, those willing to explore it will help develop our understanding of coral diseases, and will support scientists and management groups in making progress to protecting these dynamic ecosystems.

The information provided in this manual is relevant to many aspects of reef management. The identification of disease outbreaks and outbreak precursors, as well as monitoring and managing the overpopulation of known disease vectors are vital components of reef management that we hope will be more widely used to improve our collective understanding of coral health dynamics. There is a focus on Compromised Coral Health Surveys to collect in- depth data on the points mentioned above. This also allows for tracking the geographical distributions of diseases throughout reef ecosystems and facilitating rapid response programs by local resource managers. Each disease has a unique health code to make surveying more practical when conducting in situ data collection. Additionally, identification of coral disease is crucial for coral nursery and artificial reef work. As discussed throughout the book, diseases and other symptoms can often spread. Knowing what to avoid during coral fragment collections to side-step disease spread onto artificial reefs is vital for long term coral restoration success.

Whether you are a new found marine science enthusiast or a matured reef ecologist, you are encouraged to utilize the amazing world of the internet to access the endless information that has been made

Introduction to coral health, disease and surveys

available over the past few decades. This book addresses issues relating to coral health in the Indo-Pacific, with many examples and photographs; if you wish to take this knowledge elsewhere you will undoubtedly need to dive further into the literature available on coral diseases in other regions. Some diseases described in this book are relevant on a global scale (e.g. sedimentation damage), but others are not and will require additional research (e.g. white band disease).

1.3. Coral Disease Terminology Basics:

The term Coral Disease refers to any describable symptom of a coral polyp(s) that inhibits growth and reproduction or causes mortality. Disease causes are broad and variable. Biotic Diseases are those that have viral, fungal or bacterial causes. Abiotic Diseases are caused by the physical conditions of the environment such as temperature, predation, abrasion, competition, etc.

In biotic diseases, the bacteria, viruses, or fungi of are described as the Microbial Agent or Causal Agents. For example; a species of Vibrio bacteria is responsible for White Syndromes in the Pacific. The Vibrio bacteria is the microbial agent in this case. Microbial diseases are often complex, as they may be caused by multiple organisms, or by organisms which generally are dormant but then become activated due to environmental factors. To further complicating things, once dormant viruses may become active in corals which become stressed or resource limited. So, asymptomatic disease carriers can become symptomatic. This brings us to the terms Disease Triggers and Abiotic Controls. These are the external factors of stress that can influence the potential for a colony to express a disease, or even for a disease outbreak to occur. For corals, there are many different Disease Triggers and Abiotic controls including:
- Changes in ocean temperature
- Excess nutrients
- Sedimentation
- Physical damage or stress
- Changes in ocean chemistry
- Transmission from fish farming, agriculture, and aquaculture
- Marine and terrestrial based pollutants and toxic chemicals such as fertilizers

Biotic disease can be further separated into those which are infectious, and those which are non-infectious. Infectious diseases can be transmitted from one coral to another in many ways, such as direct contact, transmission through the water column, or transmission through an intermediary organism, known as a transmission vector. Non-infectious diseases generally, but not always refer to those brought about through changes in the abiotic conditions or stress. To use humans as an example, the common cold is caused by rhinovirus, which is transmitted from one individual to another through contact, air born particles, or the exchange of bodily fluids; and thus is an infectious disease. On the other hand, diabetes or cancer can represent non-infectious diseases in humans, as environmental conditions are usually a causal factor, so the disease lacks a pathogen to spread. Although in humans viruses spread easily through the air or on surfaces, in the ocean direct transmission

Introduction to coral health, disease and surveys

through the water column is thought to be rare, instead these diseases are spread through intermediary agents known as Transmission Vectors. Coral predators such as Drupella snails, Crown of Thorns Starfish and Butterflyfish are known to be disease transmission vectors between colonies, and animals such as sea turtles have been identified as vectors of disease transmission from one region to another.

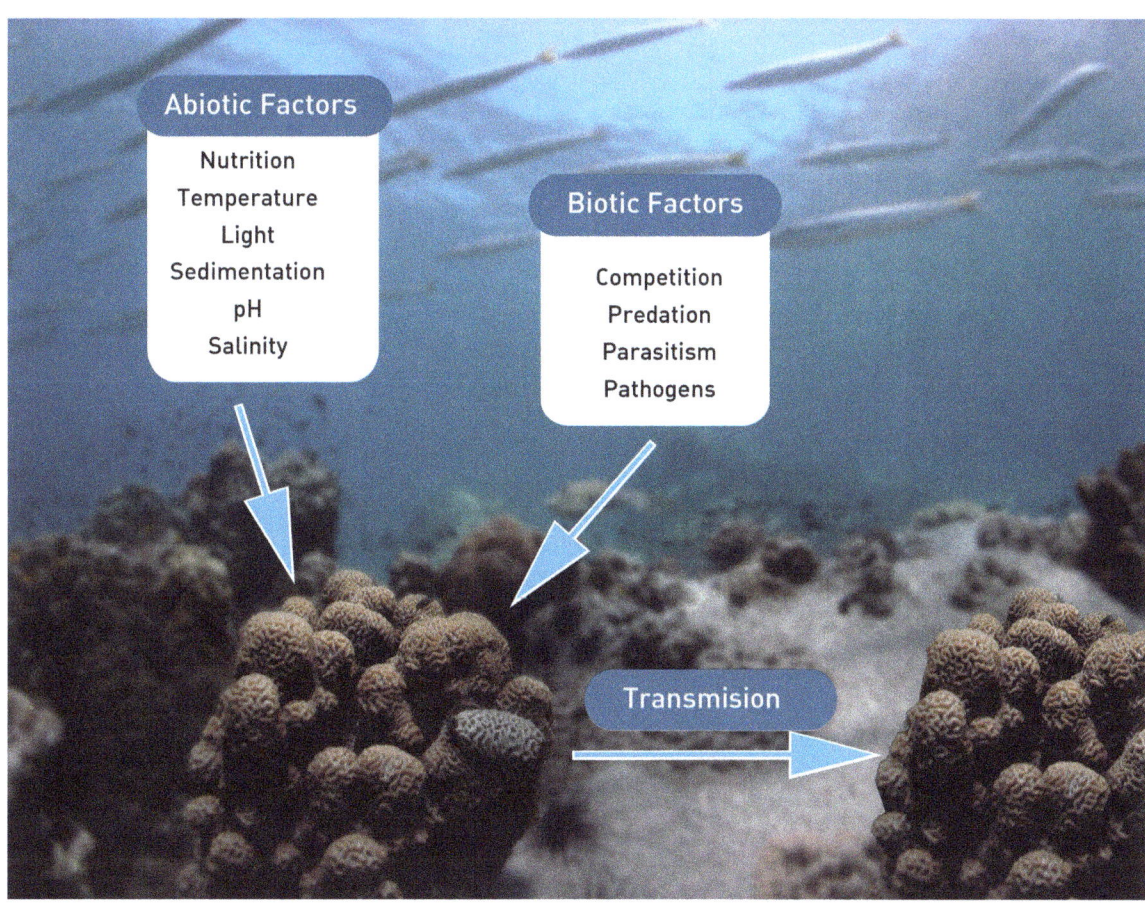

1 Introduction to coral health, disease and surveys

When discussing disease, there are several terms that are used to describe the extent or rate of spreading; Prevalence Rates, Incidence Rates and Mortality Rates. The Prevalence Rate refers to the number of individuals affected at any one time and is often described as a percentage from this equation:

$$\frac{\text{Number of affected individuals in the population}}{\text{Total number of individuals in the population}} \times 100$$

The Incidence Rate is the amount of new infections over a given period of time. It is often described as a percentage from this equation:

$$\frac{\text{Number of new infections over the time period}}{\text{Size of the at risk population over the time period}}$$

The Mortality Rate is the number of individuals that die over a time period in a given population. It is often described as a percentage from the equation:

$$\frac{\text{Number of deaths}}{\text{Total population size}} \times \frac{1}{\text{Time period}}$$

> **Did You Know?**
> Diseases with similar characteristics are not uncommon and there are discrepancies between symptoms. A study conducted by Linop et al. (2008) displayed frequent identification inconsistencies in the field of coral diseases, even amongst experts. In an emerging science like coral diseases, scientific consensus can change quite rapidly, so 100% accuracy is not required in this field for effective monitoring and management. However, keeping a comprehensive record of disease presence, prevalence and mortality is hugely important for managers even if they are only just beginning to document coral fitness. In the absence of understanding, a camera can be a crucial asset that can be used as a reference in the study of coral pathology.

When describing coral diseases, we use specific terms to refer to the observed disease symptoms. The first term used refers to the damaged areas of the coral, which are known as lesions. Lesions are areas on corals that have experienced loss of tissue, discoloration, swelling, or other physical damage. We can then describe the lesions using several terms that are accepted across the community of disease scientists. The first is the pattern of lesions, which is divided into three categories:

Introduction to coral health, disease and surveys

Lesion paterns
Lesions form various patterns on a colony. They are described in three ways:
- **Focal:** The lesion begins at a single point and progresses outward.
- **Multifocal:** The lesions start at multiple, yet distinct, spots.
- **Diffuse:** Describes lesions which are not easy to differentiate between, or the lesion is not clearly defined.

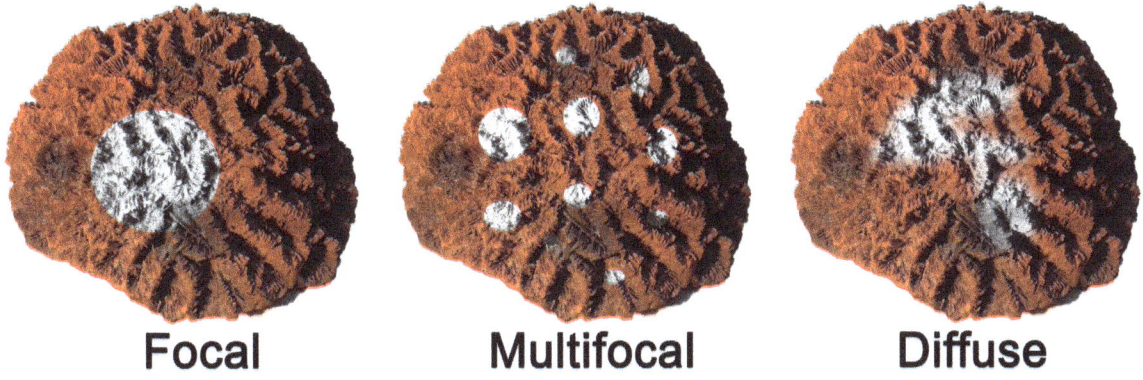

Lesion/Band shape
Next, we describe the shape of the lesion, which again is broken into three categories:

- **Linear:** The lesion moves across a colony in a relatively straight, belt-like formation.
  Note: The term 'band' for linear lesions is also commonplace amongst disease related literature and in the field. It is a term which is also used in the names of some diseases, e.g. Brown Band Disease.
- **Annular:** The lesion radiates out from a central point.
- **Irregular:** The lesion spreads randomly or has no clearly defined shape.

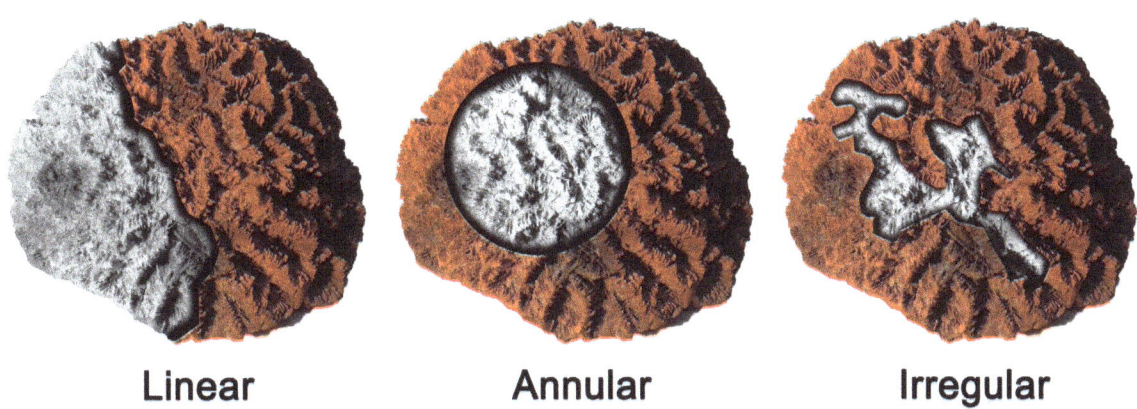

Introduction to coral health, disease and surveys

Margin borders

Margin borders (edge of lesion) are described in two ways:
- **Discrete:** The line or band bordering the living to dead tissue is obvious, defined and may have colouration.
- **Diffuse:** There is no obvious margin between the living tissue and coral skeleton, and the lesion has a diffuse gradient present from living coral to skeleton.

Rate of progression

The rate of progression allows us to estimate how quickly a disease is progressing, if at all. It is described in three ways:
- **Rapid/Acute:** The disease is spreading quickly. It can be identified by a clean white skeleton caused by recent tissue loss.
- **Moderate/Subacute:** The disease is spreading at a slow to moderate rate. A diffuse gradient of fouling organisms are present and the area around the lesion border is cleaner than the area further away from the margin.
- **Not Progressing/Chronic:** The disease has stopped progressing and the lesion area is covered in a dark layer of filamentous algae and/or fouling organisms.

Introduction to coral health, disease and surveys

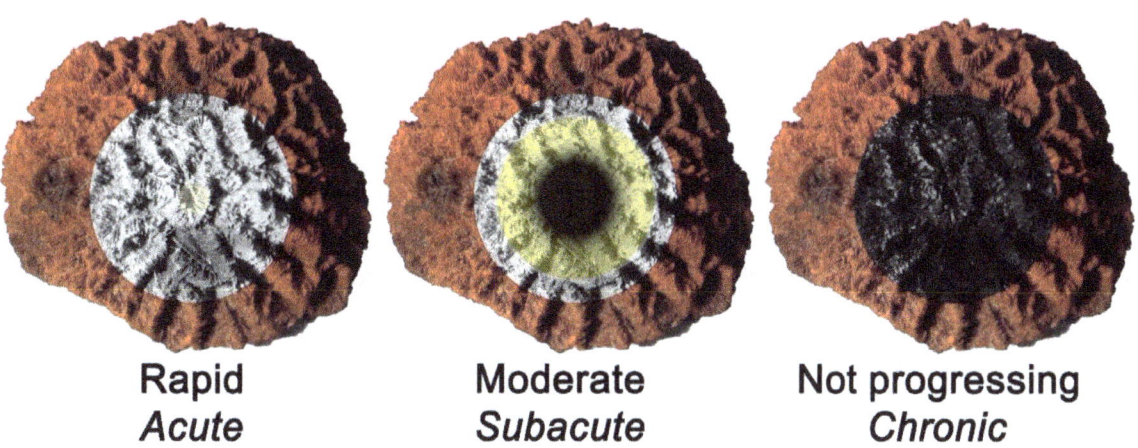

Rapid	Moderate	Not progressing
Acute	*Subacute*	*Chronic*

To achieve incidence rates, mortality rates and prevalence rates we must of course include all coral colonies found during surveys. So, even though this book focuses around being able to identify corals in compromised states of health, it is also vital to identity completely healthy colonies. Our first state of health is therefore:

Healthy (Code: H)

Indicative of a completely healthy colony, the tissue is intact and displays a dark and consistent colour, representing a healthy symbiotic relationship with zooxanthellae. If other issues are present such as overgrowth or a disease, remember to note this.

Next, let's discuss the different causes of coral mortality, their background information and how to identify each. For simplicity they are divided into four categories: Predation, Overgrowth, Disease (in the typical bacterial, fungal and viral sense) and 'Others'. Each is explained in detail, and includes a shortened code for each to be used during surveys. We will also divide the coral disease chapters into two; one focusing on common coral diseases of the Indo-Pacific and the other focusing on the common coral diseases of the Caribbean.

Chapter 1 Review

After completing the reading and discussion of the material covered in Chapter1, you should understand and be able to answer the following questions.
Please talk with your instructor about any questions you may have.
- What are the two fundamental types of coral diseases?
- How can we find the incidence rate of a coral disease?
- List three types of lesion patterns and describe them.
- Name the two types of margin borders and describe them.
- At what speeds can coral diseases progress?

Chapter 2: Predation

A healthy *Hydnophora sp.* coral colony. Photograph: Elouise Haskin

Coral predators are an integral part of a healthy coral reef ecosystem, and corals have co-evolved alongside these predators and related organisms to form the intricate network we refer to as the reef ecosystem. They come in varying forms and can be split into two main categories: obligate corallivores and opportunistic corallivores. Obligate corallivores are predators that feed on corals out of necessity, as a core component of their diet. Opportunistic corallivores are those that may feed on corals occasionally depending on availability, however do not depend on coral as a crucial component of their diets. A spectrum of corallivory exists and in addition to obligate and opportunistic coral feeders, predators which inadvertently kill corals also inhabit reef ecosystems. This category consists of marine animals which fortuitously kill polyps during the acquisition of their food, but do not directly feed on the corals themselves (see triggerfish below).

For most of history, coral predators have existed in relative equilibrium with their coral prey. For coral reefs to become the productive and diverse habitats that they are, they have always needed to recover and grow faster than it can be eaten and destroyed. Ecosystems become more resilient in the face of small disturbances, as long as there is ample time for recovery between disturbances. In the current period of heightened anthropogenic stress, some coral predators are able to take advantage of stressed corals and consume them faster than they are able to regrow.

Let's look at an example of how coral predation can become unbalanced for the ecosystem. Nutrient loading from untreated sewage can lead to blooms in phytoplankton, a source of food for the larvae of many reef organisms. With an abundance of food, more larvae are able to get the energy they need and survive to the metamorphosis stage. Normally, most would starve to death or not have enough energy to reach the next like stage, thus limiting their population. With these limits removed, populations of coral predators such as snails and sea strs can increase rapidly, to what is known as an overpopulation or outbreak.

This is the cumulative effects of anthropogenic stress to benefit certain marine organisms, often to the detriment of others. The nutrient pollution helped the predator populations, while stressing out the coral populations, making them more readily consumable. Under these destabilized conditions coral predators have the potential to cause catastrophic declines to coral coverage on reefs. One example includes the Crown of Thorns Starfish (COTS) which has had multiple outbreaks on the Great Barrier Reef in Australia, causing significant coral mortality since the 1960's. More information on these outbreaks can be found below and in the Conservation Diver EMP manual.

Common coral predators which should be monitored during reef surveys due to their capacity to alter coral coverage and population structure are discussed in the following pages.

Echinodermata

Animals from the phylum Echinodermata live within marine environments and are characterized by a variety of morphological characteristics including: radial symmetry, a water vascular system, tube feet and an endoskeleton. Some common examples of echinoderms include sea cucumbers, sea stars and sea urchins. Below we will discuss corallivorous echinoderms.

2.1. Crown of Thorns Sea Star (COTS) Code: COT

There are eight species of COTS endemic to the Indo-Pacific that are among scleractinian coral's most effective predators. Their range stretches from the Red Sea and East coast of Africa, to the Pacific Islands and the Western coast of Central America. They are not found in the Atlantic.

The Crown of Thorns can reach sizes of more than 50 cm in diameter, making them some of the largest natural predators that corals have to contend with. The sea star have earned their common name from the venomous spines or 'thorns' that cover their dorsal side. These spines and the rest of the COTS's body contain a toxin that the sea star

2 Predation

Crown of thorns sea star. Photograph: Chad Scott

is well known for called saponin. This is a haemolytic that can cause pain, swelling and even paralysis in extreme cases. As natives to the Indo-Pacific, they have played vital roles on reef ecosystems by removing sick, weak, or dying corals from the gene pool, thus strengthening the population's overall genetic fitness.

COTS play a role in that same way as other extensively studied predators. Take for example, the relationship between lions and zebras. The lions are unable to catch the fastest individuals in the herd, and thus go after the sick, injured, or young zebras. This in turn strengthens the overall population by eliminating individuals that are sick, injured, or old, and thus removes those animals from the breeding populations and in some cases,

A COTS feeding on a Pavona sp. coral, leaving a clean scar behind. Below is an older scar, collecting fouling organism growth and appearing yellow in colour. Photograph: Elouise Haskin

Predation

Examples of the scars left behind by COTS on *Acropora sp.*. Photograph: Elouise Haskin

prevents the further spread of disease. In areas such as America, where these predators are removed, herds of elk and deer have become infected with Chronic Wasting Disease, which is now a major source of mortality in the absence of natural predators. Similarly, the COTS can detect corals which are sick or stressed using chemical cues, and thus remove those weakened individuals from the local population and open up space for the settlement of new corals, helping to further halt the spread of disease.

COTs selectively eat certain species of coral. This prey preference can have important implications for their impact on coral populations. This preference is often specific to particular regions and can vary depending on the local coral population diversity and abundance. COTS indigenous to one region might show a preference to particular coral species or genera, and often leave the rest. Frequently, corals consumed by COTS include those that are fast growing and devote less energy to defence, such as Acropora spp. and Montipora spp.. They may also go after coral with thick tissues such as those of the family Fungiidae. Structurally complex and slower growing corals are less likely to be the preferred species for COTS as they are better able to protect themselves, and provide little food in the process.

Over- fishing and over-extraction of natural predators or nutrient influxes both can encourage COTS overpopulation.

2 Predation

> A significant example of COTS outbreaks originates from the Great Barrier Reef (GBR) in Australia- a reef system that has previously seen 50% coral decline between 1985 and 2012. Estimates claim 42% of this has been as a result of COTS outbreaks. The first outbreak was documented in the 1960's. Since, they have suffered from three additional plagues that were recorded paralleling major river run offs that border agricultural regions, supporting the suggestion that land-based nutrient excess assisting outbreaks. So far, GBR management groups have performed culls to reduce coral loss, and to date over 500,000 COTS have been killed. Research into population dynamics and outbreaks are continually sought by management authorities to aid in coping with these prevailing predators.

Additionally, pollution, bleaching events, and other coral stressors have become more frequent, causing coral colonies to release stress signals more regularly. Not all reefs around the Indo-Pacific are experiencing problems with COTS, however they still hold the potential to alter coral population dynamics, and so should be monitored when possible. Furthermore, every reef will have different carrying capacities for coral predators, and this can even change seasonally on a local reef. Ecosystems with low coral coverage have lower COTS carrying capacities, as even a small number of sea stars could have a potential to alter the coral community. Reefs with high coral coverage and high diversity will be able to tolerate higher populations of COTS.

Identifying COTS scars is relatively easy. COTS do not have teeth or radula. Instead, they egest their stomach onto their prey and secrete enzymes to break down the coral tissue before absorbing it through their digestive tract. Due to this feeding method, feeding scars can be either discrete or diffuse, and are likely to be rapid in regards to their rate of progression. If the opportunity to feed continuously is available, they will leave a patch of clean coral skeleton with a distinct border to the surrounding tissue. In this situation the COTS will often be nearby, hiding. The size of the scar generally depends on the size of the starfish and, if uninterrupted large amounts of coral tissue can be consumed. As time progresses, the exposed coral skeleton will become colonized by filamentous algae and other organisms. Depending on the time spent feeding and the scar size, it may become colonized by fouling organisms evenly, or form a gradient.

Gastropoda

Gastropods are a class of animals from the Mollusca phylum. The etymology comes from the Greek words gastro meaning 'foot,' and poda meaning 'foot,' and describes the key morphological traits of these animals. Gastropods encompass all snails and slugs, including all fresh water, marine and terrestrial species. Below, we discuss corallivores gastropods.

2.2. Coralliophila Snails - Code: COR

Coralliophila is a genus of sea snails which contains over 100 species. Geographically, Coralliophila are largely distributed throughout tropical to subtropical zones and are frequently reef-dependent. Their overall distribution is not reflective for all species within the genus, as many are confined to specific geographic regions. For example, the species Coralliophila abbreviata is known as one of the most important gastropods for monitoring in the Western Atlantic and Caribbean. Shells grow up to 4cm in length, form a smooth murex shape (with variability between species) and may be overgrown by crustose coralline algae (CCA). The snails are most active at night and are so far known to prey on a wide variety of Anthozoa, including Scleractinia, Coralliomorpharia and Zoantharia (Potkamp et al. 2017).

When feeding on Scleractinia coral, Coralliophila snails leave focal scars with discrete edges and are often found in small groups of clustered individuals. They feed using a radula, a feature which can be compared to a calcified tongue. It is a chitinous and toothed device used to scrape away at a food source. The genus has been recorded feeding on a variety of growth forms of Scleractinia and have a preference for the coral margins where living tissue meets skeleton. Studies have suggested that in some cases, corals are able to heal from the inflicted scars within a matter of weeks by regrowing lost polyps. However, in approximately a third of cases, scars have been noted to develop into white syndrome like lesions, and progress into larger moderately progressing focal scars (Raymundo and Weil 2016). The easiest way to tell if a coral lesion has been caused by Coralliophila snails is to check the lesion's immediate surroundings.

2.3. Wentletrap Snails - Code: WEN

Wentletrap snails, or, Epitoniids (Epitoniidae: family) are small ectoparasites (those that live on the outside of their hosts) that are known to have associations with Anthozoans such as Zoantharia, Actiniaria and Scleractinia. Some common Scleractinia hosts include stony corals from the families; Dendrophylliidae, and solitary individuals from the Fungiidae family. They vary in size from just a few millimetres in length to up to five centimetres and have elongated, conical shells. Their name 'wentletrap' is Dutch, which translates to spiral staircase, given to the gastropod because of the shape to its shell. The snails are a ubiquitously distributed family, inhabiting tropical to cold water habitats, however this manual will focus on

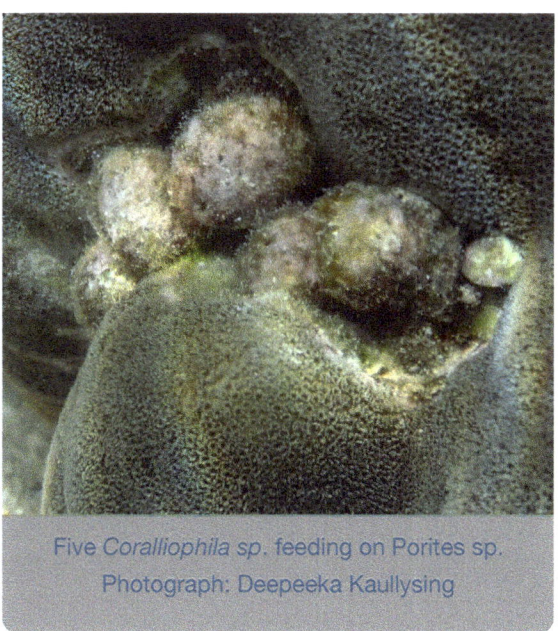

Five *Coralliophila sp.* feeding on Porites sp.
Photograph: Deepeeka Kaullysing

Predation

Epitoniids from the Indo-Pacific and Indo West Pacific that impact coral population dynamics.

Corals from the Fungiidae family are suggested to be a preferred host group for Epitoniids in the tropics, specifically, the free-living individuals which typically have an inverted lower side, providing protection from the external marine environment. Under these corals is often where the Wentletrap snails and their spawn are frequently found. On Dendrophylliidae corals, often nicknamed sun corals or tube corals, Wentletrap snails occur more so out in the open, and can be seen feeding on their tissue, and leaving spawn over the tops and edges of the corallites. It can be difficult to positively identify the feeding scars of these snails, however much like with other corallivores

An Epitoniidae feeding on a Dendrophyllidae. Photo: Rahul Mehrotra

gastropods, the presence of the ectoparasites in the immediate surroundings is the simplest way to make an accurate ID of what killed the coral colony.

Two Epitoniid snails (both centre) lays egg bundles on a *Turbinaria sp.* colony, amongst the sedimentation. Photograph: Spencer Arnold

A Dendrophyllia aureum (centre of photo, below coral polyp) predates and lays it's egg bundles (above and to the left) on a Tubastraea sp. colony. Photograph: Spencer Arnold

2.4. Drupella Snails - Code: DRUP

Drupella are native to the Indo-Pacific, and so far have not been reported as problematic in the Atlantic Ocean. They play an important role as corallivores, feeding on sick and stressed coral colonies. They are small muricid (Murididae: Family) snails that grow to a maximum of five centimetres. Their natural colouration and patterning vary between whites and browns, however as they grow, a thin layer of CCA forms over their shell, resulting in a pink to purple colouration. They scrape and consume coral tissue with a radula, a specialised mouth which is comprised of multiple plate-like structures. Similar to many invertebrate species, Drupella snails have an R-selective reproductive strategy, putting most of their energy into producing thousands of gametes for spawning events, and thereafter invest no energy into parental care of their offspring. This strategy enables them to take advantage of unstable ecosystems and fluctuating resource availability.

There are six species of Drupella in total and, for a long time, the snails went overlooked. That is until they began appearing in outbreak proportions in Japan in the 1980's. They are efficient coral predators, irrespective of their small size and what they lack for in speed they make up for in sheer numbers. From the mid 1980's to early 1990's, Armstrong SJ (2009) recorded Drupella cornus on a 100km area of the Ningaloo Reef in Australia in higher abundances than anywhere else in the world. Along some patches of this stretch, the Drupella were found to be responsible for near 100% coral

2 Predation

Drupella sp. clustering on an *Acropora sp.* colony, feeding on its tissue.
Photograph: Elouise Haskin

mortality. There is relatively little known about this small gastropod, but it is thought that they are able to live between 25-45 years.

Typically, Drupella snails bundle together in patches amongst branching Acroporids and leave clean skeleton behind. Additionally, they have also been recorded feeding on other genera including Platygyra, Pocillopora, and the Faviids family and are considered to be transmission vectors of coral diseases. Drupella scars create focal or multifocal lesion patterns with distinct boundaries. The lesion edges are where Drupella are most likely to be found. Due to the cryptic nature of the animals, in low populations they are not necessarily easy to spot. During periods of outbreaks, Drupella can occur in such high abundances that the coral substrate which they are predating on isn't visible underneath their bodies. Thus, in the case of an outbreak, this makes lesion cause identification relatively straight forward. Keep in mind that for all coral predation, bare skeleton left behind will accumulate fouling organisms, making identification more difficult over time, and other causes of mortality may need to be considered.

Drupella snails have been highly successful in the Gulf of Thailand, on Koh Tao. After the global bleaching event of 2010 Drupella were recorded on 48 genera of corals, displaying a high level of dietary plasticity. Furthermore, we (Hoeksema, 2010) identified that they went through a feeding shift, switching from their preferred prey species to the widely available mushrooms corals. Since 2010, the populations have remained high. Local efforts to control the outbreak have been regularly conducted by Conservation Diver and its partner organizations to ease the resulting coral mortality rates.

2.5. Damselfish - Code: DAMS

Damselfish (Pomacentridae: family) are common, and will be found during surveys from any tropical region around the globe. They are morphologically variable, fluctuating from 1-35 cm, and come in a profusion of colours and patterns. There is around 300 species worldwide, and are most abundant on reef ecosystems. They are algal feeders, yet frequently cause coral mortality on reef ecosystems by smothering patches of coral tissue in order to have an area to cultivate their preferred algae species on the skeleton. From there, they maintain and defend their algae patch so it can proliferate.

A Damselfish (*Stegastes sp.*) protecting its patch of algae.
Photograph: Elouise Haskin

2 Predation

Predation lesions by Damselfish are generally focal with a discrete edge. Since Damselfish 'farm' algae, the lesion can appear as though it is non-progressing, or chronic. Many novice Conservation Diver students mistake this form of coral mortality as macro algae overgrowth (discussed in chapter 4). It can be useful to keep in mind that Damselfish are territorial, and frequently defend their algae patch aggressively to fend off other algae feeders. Subsequently the presence of an algae patch with a defensive Damselfish is a practical way to identify the cause of coral mortality.

2.6. Butterflyfish - Code: BTFY

From the family Chaetodontidae, Butterflyfishes are found in tropical waters around the equator, and can be found in all tropical oceans. They are brightly coloured and have pinched rostrams, or mouths, to assist in their varied diets of small invertebrates, algae and coral. There are around 130 species. Many butterflyfish include coral polyps in their diet, with some; the Eight Banded Butterflyfish (Chaetodon octofasciatus), for example, being an obligate corallivore.

An Eight Banded Butterflyfish predating on a *Diploastrea heliopora* coral. Often the scars are easily mistakeable with White Syndrome Disease. Photo: Elouise Haskin

Predation

An Eight-banded Butterflyfish hiding amongst an *Acropora* coral colony. It will use the colony both as habitat, and as food. Photograph: Elouise Haskin

Three types of Butterflyfish (from left to right): Weibels Butterflyfish (*Chaetodon wiebeli*), Longfin Bannerfish (*Heniochus acuminatus*) and the Copper Banded Butterflyfish (*Chelmon rostratus*). All three include coral as part of their diet. Photograph: Elouise Haskin

2. Predation

Butterflyfish are recognized as transmission vectors of coral disease. A study by Raymundo et al. (2009) conducted in the Philippines looked into how trophic diversity of fish communities influenced coral disease abundance across both Marine Protected Areas (MPAs) and non-MPAs. They found increases in fishes from the Chaetodontidae family positively correlated with increased coral disease abundance.

Butterfly fish lesions have a moderate progression, reflecting their slow feeding rate and are focal or multifocal with discrete borders. Consequently, if the lesion is large enough, a feathered progression of filamentous algae will build up. They are easiest to identify if Butterflyfish are present and feeding. If the fish are not feeding, checking the borders of lesions and ensuring the scar isn't due to another type of disease is necessary. Be careful as you inspect the lesion, as White Syndrome looks similar to Butterflyfish predation, remembering that Butterflyfish leave a discrete border, while the border of White Syndrome may slough off the coral skeleton at the disease border (discussed in chapter 7).

2.7. Triggerfish - Code: TRIG

Triggerfish (Balistidae: family) are a group of around 42 that live within tropical to subtropical shallow reef ecosystems or rocky outcrops, with one species, the Canthidermis maculate living in pelagic waters.

Their diets consist largely of sessile or slow-moving Molluscs, as well as Echinoderms and Crustaceans. They are able to extract boring molluscs, worms, and other small animals hiding within coral colonies by breaking through coral skeleton with their teeth. Thus, the coral lesions left behind by triggerfish will look like broken off 'tops' of coral colony peaks, and may have oval-shaped excavations within the lesion (this excavation being the previous site of the predated mollusc). In other cases, they will appear like teeth scraping inwards towards the prey location. The lesions are unlike that of other coral predator lesions, however they are mostly focal or multifocal (depending on the growth form of the coral), and the skeleton will be exposed suddenly as a result of the predation. Triggerfish are common in the Indo-Pacific. The Gulf of Thailand for example, has high abundances of the Titan Triggerfish (Balistoides viridescens) and the Yellow Margin Triggerfish (Pseudobalistes flavimarginatus) and so, triggerfish predation scars are common in their waters, especially in corals with boring invertebrate molluscs

The scars of a Triggerfish on a *Porites sp.* coral. Photograph: Pau Urgell Plaza

and worms. The largest Triggerfish species (Balistoides viridescens) can grow up to about 75 cm long, however there are many smaller species which will vary from 25-50 cm. They have large front teeth which are easy to spot from front on.

Much like the damselfish, triggerfish cause coral mortality merely to access a different food source. Even if the lesion is large, the breaking of the coral happens in a moment and clean white skeleton is exposed below. Similarly to other scars, with time the white will be replaced with fouling organisms, and in some cases another disease may develop around the lesion due to the stress of the fractured coral.

A *Scarus rivulatus* grazing for macroalgae.
Photograph: Spencer Arnold

2.8. Parrotfish -Code:PRT

Parrotfish belong to the Scaridae family of fish, with around 80 different species ranging in size from just a few centimetres during their juvenile phase to over one meter in length, the largest being the Bumphead parrotfish (Bolbometopon muricatum). There is a relatively consistent elongate oval body shape amongst most species, yet colours and patterns vary greatly. One of the key identifiable characteristics of parrotfish are their beak-like mouths, used in scraping at their preferred food types. They are predominantly algae feeders and their 'beaks' allow them to scrape the filamentous or macro-algae off of surfaces. However, they may occasionally consume living corals to feed on unicellular algae within the coral tissue. This scraping technique allows Parrotfish feeding scars to be easily identifiable; they are multifocal and often appear in small ovals with damage to the skeleton below the missing tissue, due to the scraping nature of the predation. Generally, they will also be on the tops, or horizontal surface of the coral where the tissue is the most thick. Whilst the initial impact of the scrape makes progression rapid, there have been many accounts of coral colonies healing from parrotfish mortality.

2 Predation

A *Porites spp.* colony with shallow scratch marks as a result of Parrotfish predation. The skeletal structure within these scratch marks has been damaged.
Photograph: Pau Urgell

Chapter 2 Review

After completing the reading and discussion of the material covered in Chapter 2, you should understand and be able to answer the following questions.

Please talk with your instructor about any questions you may have.

- What is the major difference between a feeding scar and disease progression?
- Why are corallivores important on reef ecosystems?
- What factors lead to corallivore's ability to become problematic on reef ecosystems?
- When would we consider predators to be overpopulated?
- Which coral predators are obligatory corallivores?

Chapter 3: Competition

Competition is a constant on coral reef ecosystems. When we observe a healthy reef ecosystem, there are countless examples of sessile organisms trying to outgrow one another. Epibiosis is a term that describes the competition between sessile organisms which colonize the surface of living or non-living substrates. The phenomenon is frequently observed on coral reef ecosystems as stable substrate is both a valuable and limited resource. Healthy reefs have high coverage and diversity of hard corals, which encourages the species diversity of non-coral organisms. For example, fish may use particular corals as juvenile nursery grounds, or as a space to avoid predation during sleep, or even as a food source, depending on their species. These species may have vastly different ecological niches and roles. Herbivores and meso-predators, help maintain flora and fauna in lower trophic levels when their abundance is in balance with their food sources. They then are controlled by the top predators who feed on them. In all cases, a role is played that maintains the health of other organisms, which carry on playing their own ecological roles, increasing the overall health and resilience of the ecosystem. The impact of the removal, or out competing of a single important species can lead to altered trophic cascades and may have lasting effects on the ecosystem.

Some cases of overgrowth can indicate a variety of changes or disruptions on a coral reef ecosystem. One example, which will be discussed in further detail later in this chapter, is sponge overgrowth. Sponges are filter feeders that pull food and nutrients from the water. Some colonies are able to filter thousands of litres of water per day, and thus reduce the abundance of nutrients in the waters of the reef. Under some conditions, such as high nutrient densities, these organisms may out-compete nearby corals. It is important to monitor this type of mortality in corals, as it tells us a lot about how local ecosystems are handling environmental changes. This chapter explores competition and the types of overgrowth we see frequently in the waters of the all the Earth's oceans.

3.1 Macroalgae/Nutrient Indicator Algae Code: MA

Macroalgae are an important part of many marine ecosystems, such as sea grass beds. However, they are not a major component of a healthy reef ecosystem. As such, they are useful indicators of nutrient excess on reef ecosystems. There are three main phyla of macroalgae that live on coral reefs, including green algae (*Chlorophyta*), red algae (*Rhodophyta*) and brown algae (*Heterokontophyta* or *Ochrophyta*). Red algae are the most diverse group of macroalgae existing on coral reefs and include the crustose coralline algae, a reef builder which provides a vital role on coral reefs by re-stabilising damaged reefs, and in some cases creating physical barriers to wave action and other disturbances. Brown and green macroalgae are more frequently indicators of excess nutrients on coral reefs and can also indicate other environmental changes, one of the reasons why these marine plants are

3 Competition

so frequently used as bioindicators. Corals evolved under oligotrophic conditions (nutrient poor), and feed on organic detritus in the water column, but mostly rely on photosynthetic energy. When nutrient levels increase, many species of macroalgae utilize it and can become over-abundant, outcompeting hard corals and other dominant reef-building species. If nutrients continue to increases further or an additional stress is added, there may be a phase-shift to an algal dominated ecosystem. Most macroalgae are easily identifiable as multicellular plant-like structures on the reef, often utilising rubble spaces to grow. They don't create visible lesions as other forms of coral mortality do; however, their overgrowth can progress rapidly when conditions are favourable.

Macro Algae *Padina sp.* overgrowth on an Acropora coral colony. Photograph: pau urgell

3.2. Cyanobacteria/Filamentous Mats
Code: CB

Cyanobacteria growing over a *Porites* coral, it is progressing annularly. Photograph: Elouise Haskin

Cyanobacteria are ancient organisms whose ancestors were the first to carry out photosynthesis, providing the basis of our modern day food web and creating an oxygen-rich atmosphere. Consequently, it drove the great oxygenation event over two billion years ago, killing off many anaerobic respirators. While cyanobacteria gave rise to the oxygen rich world we know; it also plays a role on reef ecosystems as an occasional competitor or fouling organism of corals.

Cyanobacteria, often referred to as blue-green algae, is comprised of a consortium of bacteria and other microscopic organisms that seasonally settle to create cyanobacterial blooms in the form of mats, clumps, and colonies. If organic matter exists where a cyanobacteria bloom settles, it is likely that the organism will be outcompeted and smothered. It is hypothesised that wave action, winds and currents prevent blooms. Consequently, calmer and warmer waters

Competition

A cyanobacteria bloom over a field of Pocillopora corals. Cyanobacteria can settle over nearly any type of substrate. Photograph: Spencer Arnold

are suggested to be optimal for their development. They vary from dark greens, reds, purples and grey/blacks. Many cyanobacterial clumps are fragile enough that they disintegrate into the water column when disturbed, greatly aiding their dispersal and colonization of coral reefs. Cyanobacterial matting can occur rapidly when the conditions are favourable.

Cyanobacterial blooms vary in progression speed, ranging from rapid to not progressing, depending on conditions. Shapes of the lesions when overgrowing corals are also a spectrum, and cyanobacteria coverage can be multifocal or annular or linear. When identifying cyanobacteria, remember that the mats will frequently disintegrate if disturbed. If you are unsure if your ID is correct, you can carefully wave your hand near the affected area and see if the mat persists or breaks apart in the water column.

3 Competition

A Didemnum molle colony growing over a dead Fungiid coral. Photograph: Elouise Haskin

3.3. Tunicates/Ascidian Overgrowth
Code: TA

Tunicate is a common name given to organisms of the Tunicata subphylum, which includes Ascidians (class: Acidiacea). Their tough exterior acts like a 'tunic' made from a cellulose-like substance and encloses the animal for protection. During the adult portion of their lifecycle, tunicates exhibit the tell-tale invertebrate characteristic of having no backbone. However, in their free-swimming larval stage, tunicates contain a notochord. This notochord is later metabolised as the animal metamorphosizes into its adult stage. This unusual metamorphosis makes tunicates protovertebrates; a distant evolutionary cousin of all modern-day vertebrates.

Tunicates vary in size, colouration and morphology, ranging from orange, green, pink, transparent and more. Occurring throughout the world's oceans, tunicates have adapted to benthic, pelagic and deep-sea habitats, thriving in both temperate and tropical regions. They are exclusively filter feeders (with some rare deep water carnivorous exceptions which are outside of the

Fun Fact: Tunicata also includes two other classes; Thaliacea and Appendicularia which are also encased within a structural tunic. These two classes differ from the Acidiacea as they remain within the pelagic zone for the duration of their lives, they are frequently referred to as Salps.

Competition

Colonial tunicates overgrowing a Pocillopora sp. coral. Some colonial tunicates such are easily mistakable with sponge. Photograph: Elouise Haskin

Acidiacea class), sucking in water through their incurrent syphon, filtering out small particles such as phyto and zooplankton before excreting the remaining water through their excurrent syphon. The general shape of an individual tunicate could be likened to an egg, but can vary depending on species and physical conditions. Solitary individuals that are fastened to the substrate can range from just a few millimetres to around 7cm in height. Colonial tunicates appear as an encrusting sheath which may have only a couple of millimetres of thickness, and on close inspection the incurrent syphons are usually visible.

Historically, it has been suggested that Tunicates prefer stable substrata, however preliminary research conducted by Conservation Diver has suggested that we have much to learn about substrate preference in the modern marine environment in regards to this organism. Information is generally lacking on this puzzling organism, especially regarding population outbreaks. For example, Didenmnum molle is from the Acidiecea class, and early research from the Gulf of Thailand has shown it to be highly competitive for substrate, including growing over scleractinian corals. When in contact with hard corals they cause a stress response due to toxins in their tentacles, and induce bleaching over the area of contact. The harm done to corals may not necessarily cause immediate tissue loss, however the bleached lesion may spread, progressing slowly, or may with time cause tissue loss, this will depend on tunicate species and reproduction rates. Tunicates cause focal or multifocal lesions on a coral. High levels of Tunicate epibiosis on coral reefs often represents an increase in nutrient load on the ecosystem.

Three images showing Ascidian colony diversity. Photographs: Elouise Haskin

3. Competition

A Pink Sponge beginning to overgrow a *Porites sp.* coral. Sponges are easily recognisable by their excurrent syphons. Photograph: Elouise Haskin

3.4. Sponge Overgrowth - Code: SP

Sponges were the first multi-cellular animals to have evolved on earth, and have been in our oceans for over 600 million years. They are filter-feeding, colonial animals which have retained their primitive evolutionary strategy, collecting the nutrients required for growth and reproduction by filtering water through their small and numerous incurrent canals. After acquiring nutrients and oxygen, they expel excess water and waste back out into the water column via their 'chimney(s)'. Sponges vary significantly in growth form and colour, making taxonomy complicated and identification to a species level difficult. Their pores can be challenging to identify in some species, and often requires close investigation to ensure correct identification. On reef ecosystems, filter feeding organisms are valuable as they have the capacity to remove or cycle excess nutrients from the water, inhibiting the growth of other fouling organisms and providing biologically available nutrients to other reef organisms.

In some instances, like nutrient enrichment and sedimentation for example, sponges become problematic on coral reef ecosystems, competing for resources and space. While some species grow large and have the capacity to filter thousands of litres of water on a daily basis, benefiting many reef organisms, others are considered fouling organisms, competing for space and overgrowing corals. They have chemical defences protecting them from predation, while also inducing stress in nearby coral colonies, further inhibiting their productivity. If they overgrow corals, the area covered in sponge is unable to recover. Some coral species show more resilience towards sponges and

Competition

Sponge overgrowth on a branching *Porites* coral.
Photograph: Elouise Haskin

can actively dissuade the sponge from competition, although unfortunately, in many instances this is not the case. Since nutrient loading on a coral reef can benefit sponges, recording their overgrowth helps informs us about nutrient runoff in the local area. Sponge overgrowth varies from irregular patterning, banding, and linear, and can progress both slowly or acutely depending on resource availability.

3.5. Competition between Coral Colonies

Corals frequently compete for space with one another. They use their nematocyst cells to attack neighbouring colonies and can cause stress bleaching and even mortality. This competition is natural and healthy on a diverse reef ecosystem.

Two corals of the same species, *Diploastrea heliopora* competing for space.
Photograph: Elouise Haskin

Coral Disease Identification and Monitoring Manual

3 Competition

Chapter 3 Review

After completing the reading and discussion of the material covered in Chapter 4, you should understand and be able to answer the following questions. Please talk with your instructor about any questions you may have.

- Why do *Didemnum molle* cause bleaching in corals?

- What factors can cause Cyanobactera outbreaks, or 'mats'?

- How can sponges be both beneficial and harmful for reefs? Give an example of each.

- What causes overgrowth of Macroalgae? Do any other organisms in this chapter do well under the same conditions? Which ones?

Chapter 4: Other Symptoms

In this chapter, we will cover all of the other symptoms that cause disruptions in healthy coral that do not fall under the categories of predation, competition, or disease (next chapter). They are variable in cause, some being the result of physical changes and disturbances, others represent symptoms of disruption to the productivity of corals, yet we are not necessarily sure of the cause.

4.1. Unexplained Bleaching
Code: UBL

Thermally induced coral bleaching is one of the most significant threats to modern day coral reef ecosystems. The frequency and intensity of bleaching has been rising alongside greenhouse gas emissions following the industrial revolution. At the time of publication, there have been four global scale coral bleaching events; 1998, 2010, 2015 and 2016-2017. Each of these years saw reefs across the globe experience loss of the symbiotic zooxanthellae from the tissue of corals due to a rise in sea surface temperatures. Thermal bleaching appears in many different forms, as you will remember from your EMP training: Partially Bleached (PBL), Fully Bleached (FBL), Recently Killed Coral (RKC) and Dead Coral (DC).

Thermal bleaching is an important threat to monitor on coral reefs, however is not included in Compromised Coral Health Surveys. During mass bleaching events, Bleaching Surveys are conducted, that follow global protocol, established in 2010. A more comprehensive data set is collected on the intensity and type of bleaching during this survey.

Unexplained Bleaching on the other thand, is not the direct result of thermal pressure. The cause is likely to be unidentifiable while conducting in-field observations, and patterning can be focal or multifocal and borders are discrete or diffuse. It typically affects a single random colony. If you

Two examples of thermal bleaching on *Astreopora sp.* that are not recorded during Coral Disease surveys, FBL and PBL. Photographs: Elouise Haskin

4 Other symptoms

observe multiple colonies or large expanses of bleached corals, it is likely to be thermally induced. Within the Indo-Pacific, Porites and free living Fungiidae corals frequently display unexplained bleaching patterns, however all corals can potentially display this compromised heath state.

A common exa mple, Porites sp. corals are often being affected by Unexplained Bleaching (UBL). Photograph: Elouise Haskin

In addition to thermal stress and unexplained bleaching - bleaching can be induced through other means. Competition with other corals, fouling organisms, disease and even invertebrates can all induce stress, often causing an area of 'stress bleaching' which often appears in a linear fashion. Pigmentation Response is also a frequent reaction to nearby competition (refer to chapter 5). When collecting UBL data, make sure the cause is not any of these triggers.

A common example, *Danafungia sp.* expressing Unexplained Bleaching (UBL). Photograph: Elouise Haskin

4.2. Pigmentation Response Code: PGR

Pigmentation responses (PGR) are common in corals around the globe. Symptoms involve lesions of discoloured and slightly inflamed tissue. Pink lesions frequently appear on Porites spp. as hues of blue, purple, green and yellow, often dependant on the coral genus and the fluorescent proteins it contains. PGR are multifocal, annular to irregular in shape, and have diffuse border margins. Physical disruptions, such as sediment damage, invertebrate boring and fish bites can cause localised areas of pigmentation response. It is unknown whether abiotic controls play a role in PGR expression, or which microbial agents may be involved, however it has been revealed that areas affected by PGR on some corals have an increased immune response and/or reduced saturation of zooxanthellae, suggesting that the pigmentation is representing the presence of a foreign intruder (Palmer C.V. et al., 2008).

Pigmentation response is better described as a defensive response, as there is usually no tissue loss while the coral polyp(s) tries to avert off unwanted intruders. Symptoms can persist for years if the underlying causes are not addressed.

Pink Line Syndrome (PLS) is a coral disease which frequently affects Porites corals. It has a cyanobacterial microbial agent, Phormidium valderianum and is expressed on Porites corals by forming an annular to irregular pink band which spreads from a localised point on a colony. Coral tissue is usually killed from cyanobacterial smothering in cases of PLS, however it is imperative to know that like with some other diseases (discussed in chapter 5), there has historically been some confusion in disease descriptions. It has been suggested that PLS is a misnomer for PGR, which covers more general stress responses by the expression of a pigmentation response (Raymundo et al., 2005).

4.3. Growth Anomalies - Code: GA

Growth anomalies frequently affect some of the major reef building corals, and are expressed as areas of abnormal skeletal growth on a coral colony. They are sometimes tumour-like in appearance, but can describe any abnormal development of the coral tissue or skeleton. Polyps can be present, absent or sparse, and the area of abnormality can lack optimal levels of its symbiotic zooxanthellae resulting in pigmentation ranging in intensity from normal colour to white. Growth anomalies are focal or multifocal, and are circular or irregular in their protruding lesion shapes. The borders of growth anomalies are discrete. Thus far, it has not been confirmed if microbial agents play any major role in growth anomaly causes, and it so far is not known to be contagious to other members of local coral communities.

Growth anomalies can affect a coral's photosynthetic output across the affected area by around half, and therefore limit the colonies ability to grow, reproduce, create lipid stores for future periods of stress, and fulfill their other necessary somatic requirements. There is still more to learn in regards to this disease, but one study published by Aeby et al. (2011) looked into nine disease-environment associations which may contribute to the expression of GA in *Acropora* sp. and *Porites* sp. corals. The publication suggests that abiotic controls for growth anomalies included coral host population density, which

Porites sp. coral colonies are often affected by Pigmentation Response (**PGR**). Photograph: Elouise Haskin

strongly correlated with the expression of GA in both coral genera. Additionally, population density of nearby coastal communities was a strong correlating factor for the expression of GA in *Porites*. Studies like this allow for a better understanding of the causes and effects of coral disease expression, as to this day it is still a grey area that needs further exploration.

A tumour-like growth anomaly affecting an *Acropora* sp. coral. Photograph: Elouise Haskin

4.4. Sedimentation - Code: SDS

Sedimentation damage is an issue that is increasing globally due to increased deforestation and human activity. Sediment collects on the coral, reducing gas exchange and preventing the buried polyps from feeding. Corals must produce mucous to remove sediment, which also causes stress to the organism and causes them to redirect metabolic energy from other tasks, like the immune system. Prolonged periods of sedimentation can lead to shading of the corals, causing focal bleaching and mortality. Sediment also brings with it bacteria and other pathogens that can cause other coral diseases.

Abnormal growth of *Diploastrea heliopora* corallites. Photograph: Elouise Haskin

The development of coastal regions in the tropics directly impacts the adjacent reef ecosystems and seagrass beds. These marine habitats often find themselves facing the same impacts of their neighbouring terrestrial ecosystem, experiencing vegetation removal and infrastructure development. It is the excess of sediment; mud and soil from development that moves from terrestrial environments into marine environments during periods of rain. Mangrove vegetation sit in the sub-tidal zone and provide important protection from land development, by acting as a breaker for wave action and as a filter system for nutrients and sediments travelling from land to sea. When this additional barrier is removed then once again, the ecosystems sitting on the upper reef slopes have lost an additional layer of protection. As a result, terrestrial areas lacking soil stability are easily eroded, which in turn leads to excess sedimentation entering the coral reef and potentially burying or smothering corals. Sedimentation damage can occur on reefs without human interference, in natural disaster cases such as flooding or landslides that border coastal regions. For the sake of a Compromised Coral Health survey, both these instances would be recorded as sedimentation damage. Generally, sedimentation progression is slow to moderate, depending on the terrestrial circumstances.

Severe sedimentation damage burying entire coral colonies. Photograph: Elouise Haskin

During periods of heavy sedimentation on coral reefs, the immediate consequence is coral smothering. If the problem is persistent then coral colonies can be buried entirely. An additional issue arises for the local ecosystem when a reef is buried in sediment, as the area now lacks stable substrate for future coral recruits to attach to, and could prevent the reef from recovering. Monitoring sedimentation damage provides insight into the impacts of land development. It is useful for future mitigation strategies and development decisions to side step this problem.

There are examples of corals overcoming problems associated with sedimentation damage. Foliose *Pavona* colonies for example, can grow in cup-like structures allowing for the integration of sediment into their structure. The colony encrusts over and incorporates the condensed sediment with the aid of coralline algae and use it to better stabilize themselves. Since these corals are fast growing, this additional survival technique better allows them to outcompete other shallow-living genera in areas affected by sedimentation.

Severe sedimentation effecting a *Porites* coral. Photograph: Chat Scott

4 Other symptoms

4.5. Trematodiasis - Code: TMD

Trematodiasis is caused by small parasitic flatworms called trematodes (Trematoda) from the Platyhelminthes phylum. There are thousands of species, most of which affect organisms other than corals. Their method of transmission is to travel through the food chain. Successful Trematoda begin life by parasitizing a mollusc, prior to expulsion and transferal to a coral polyp. Once established in a coral host, they rely on predation by coral feeding fish, before their next generation can begin this cycle again after the excretion of their eggs in the fish hosts faeces. This method of transmission through the food chain makes the coral an intermediate host.

A *Porites* colony with Trematodes (bright pink, swollen segments). Photograph: Chad Scott

Trematodiasis symptoms persist as small pink (fluorescent) pigmented protrusions of inflamed polyps on a coral colony. These lesions are multifocal and do not grow into larger lesions. The progression, or, addition of new focal spotting is slow. It is suggested that the fluorescent proteins of the lesion are beneficial to both the trematode and coral host. The proteins attract fish predators, so the trematode life-cycle can continue. These proteins also benefit the coral colony, as the fish predator extracts the parasite from the colony, allowing regeneration of the affected area via means of asexual reproduction. While coral-embedded trematodes may perish with time if no fish predate it, other trematodes in the colony could collectively last many years (if trematode death and predation rate is met or exceeded by reproduction rate). Coral mortality is not typical in parasitized coral colonies, however growth rates of the coral can be reduced by up to 50% due to the intrusion, as affected polyps lose functionality (Aeby, 1992). So far, Trematodiasis is only known to affect *Porites spp.*.

A *Porites* colony with a variety of invertebrates boring into its tissue, including Christmas tree worms, bivalves, and Trematodes (bright pink, swollen segments). Photograph: Elouise Haskin

Other symptoms

4.6. Physical Disturbance - Code: PHY

Other physical disturbances of corals not already covered must be recorded as physical disturbances. A common example; trash, is provided below. Others may include, but are not limited to: immediate damage caused by SCUBA divers, boats causing breakage of coral in shallow bays, storm damage and dropped anchors, to name a few. Physical disturbances can be acute, such as during storms or anchor dropping, or can be chronic as in the case of improperly implemented mooring lines or ghost fishing nets. Most often, the source of the physical disturbance will still be present if not obvious from the patterns of the damage.

A *Diploastrea heliopora* coral damaged from a poorly constructed mooring point. Photograph: Chad Scott

A solitary Fungiid coral damaged from physical disturbance. Photograph: Chad Scott

4.7. Trash/Rubbish - Code: PHY, TR

This one is self-explanatory, corals being directly affected by solid waste (any item which was not intended to belong on the reef), with the exception of artificial reefs and coral nurseries should be noted here. Marine debris can cause smothering and shading of corals. In the case that the marine debris is removed, often the coral underneath will appear bleached or inflamed. Other debris, such as ropes and nets can chronically abrade the coral, and can appear as mortality.

4 Other symptoms

A plastic bag caught on a *Pocillopora sp.* colony, it could cause lesions or even spread coral diseases. Photograph: Chad Scott

Chapter 4 Review

After completing the reading and discussion of the material covered in Chapter 4, you should understand and be able to answer the following questions. Please talk with your instructor about any questions you may have.

- What terrestrial activities cause sedimentation problems on coral reefs?
- What is a Trematode?
- What are the two types of Growth Anomalies?
- Explain what pigmentation response is, and how it can express itself on a coral colony.

Chapter 5: Diseases of the Indo-Pacific

Now, we reach diseases of the fungal, bacterial and viral type. This chapter includes information on the following for each disease where it is available:

- Susceptible coral genera
- Abiotic control(s)
- Microbial communities
- Progression rates
- Disease characteristics

Globally, coral disease outbreaks have been responsible for the direct loss of corals and changes in the population structure of the corals and their associated reef organisms. Diseases are transmissible in a variety of ways. Intuitively we know that coral diseases spread most effectively in areas of high coral density and low diversity, as with other organisms. The microbial communities, or agents of coral diseases are not uncommon on reef ecosystems. It has been argued that they are opportunistic and that some can lay dormant in a coral host, and will proliferate during periods when the coral becomes stressed. Increases in disease vectors, water pollutants and high temperatures also increase the chance of disease outbreaks, likely due to reduced coral immunity or improved conditions for disease microbial communities, or a combination. The study by Lamb et al., (2014) also identified that disease incidence was positively correlated with SCUBA diving pressure. After this chapter, you will have the tools necessary to identify common coral diseases in the Indo-Pacific for regular coral health assessments. For coral disease monitoring in other oceanic regions, such as the Caribbean, additional information will be required.

5.1. White Syndrome - Code: WS

White Syndromes (WSs) are an enigmatic and prevalent group of diseases which affect scleractinian corals in the Indo-Pacific, often leading to complete colony mortality. Prior to a standardized methodological approach to describing coral lesions and their microbial communities, a variety of macroscopically similar 'white' diseases were described, some of which have been synonymised since description due to a lack of reliable diagnostic features. However, with increased knowledge some of these may be once again differentiated from WS in the future, but for now we will stick to the most commonly used nomenclature.

White Syndromes include lesions of tissue loss, exposing white skeleton with diffuse or discrete borders with no coloured band bordering the lesion margin. The affected area behind the disease front line will consist of pure white skeleton, but will darken with time as the area is colonized by opportunistic fouling organisms. Lesion shapes are focal or multi focal and can be linear, annular or irregular depending on the host colony shape. Coral tissue will slough off at the lesion margin. This group of diseases includes what has previously been described as: White Plague, Ulcerative White Spots, White Pox and White Band-like diseases. It is likely that WSs consists of multiple distinct diseases, however due to a lack of

5 Diseases of the Indo-Pacific

White Syndrome affecting a *Diploastrea heliopora* colony. From the filamentous algae build up we can see it is progressing at a moderate rate. Picture: Chad Scott

information regarding specific microbial causes, they are still frequently identified by lesion characteristics.

Confirmed components of the WS's microbial community are bacterium of the *Vibrio* genus, however studies have suggested that a large consortium is likely at play. Pollock et al., 2017, found *Rhodobacteraceae* to be a significant member of the WS's consortium. Additionally, unpigmented ciliate communities, chimera and helminths parasitism have also been found present in varying WS disease studies. A possible explanation for the large variation in WS's pathogens and epidemiology is that it may be an immunodeficiency disorder of the coral and thus, secondary fungal, bacterial, parasitic and viral components are able to infect the host after the WSs have established. This explanation can be comparable to some human diseases which affect our immune systems, such as HIV or leukaemia – where the presence of the disease leaves us susceptible to accumulating other bacteria or viruses, as host immunity has been reduced or collapsed.

The progression of WSs vary from moderate to sub-acute, depending at least partially on the host species. A high variety of coral are susceptible, with at least 15 genera in the Indo-Pacific affected by WSs. Abiotic controls contributing to the spread and infection of WSs includes increases with sea surface temperature and pollutant levels. WSs have had, and will likely continue to have a profound effect on the community structure of coral reef ecosystems.

There are at least eighteen genera of scleractinian corals in the Indo-Pacific which are affected by White Syndrome, as found by Bourne et al., 2016 and Sutherland et al., 2004, including: *Acropora*, Coscinaraea, Dipsastraea, *Echinopora, Favites, Goniastrea, Hydnophora, Leptoria, Leptoseris, Lobophyllia, Montipora, Mycedium, Platygyra, Pocillopora, Podabacia, Porites, Stylophora,* and *Turbinaria*.

White Syndrome affecting *Acropora* sp.. Picture: Elouise Haskin

5.2. Black Band Disease
Code: BBD

Black Band Disease (BBD) was the first recorded coral disease, noted in the Caribbean in the 1970's, affecting scleractinian coral, octocorals and gorgonians. Today it is one of the most significant coral diseases globally. It was first recorded in the Indo-Pacific in the mid 1980's, affecting shallow colonies, generally no deeper than 10m. It progresses linearly or annularly, rarely having more than one point of initial appearance, and consists of a dark filamentous black to dark red band, with a typical width of 1-3cm. The rate of disease progression is generally rapid, averaging 3mm per day, but some observations have noted progression as quick as 1cm per day (Richardson, 2004). Bands are discrete and may overlap some living coral tissue, and a clean white skeleton remains behind the band as progression continues, until fouling organisms settle. The abiotic controls which influence the epizootiology of BBD in the Indo-Pacific are warm temperatures exceeding

> **Did you Know?**
>
> Even though a large range of genera are known to host BBD within the Indo-Pacific, on small geographic scales often only a small handful of genera are affected, even if other confirmed genera exist in the region. For example, observations of BBD by Conservation Diver in the Gulf of Thailand only includes *Gardineroseris* and *Psammocora* as hosts.

5 Diseases of the Indo-Pacific

A *Gardineroseris* sp. affected first by thermal bleaching, then by Black Band Disease. Photo: Chad Scott

28-29 degrees Celsius, light concentrations, pollution, sedimentation damage and turbid water. The microbial agents consist of a consortium of microorganisms. Most abundant are a variety of photosynthetic filamentous cyanobacteria, including that of *Trichodesmium* sp. (Sutherland et al. 2004). Sato et al., 2010 found that in nearly a fifth of *Monitpora hispida* BBD cases studies on the Great Barrier Reef, the banded disease developed after the settlement of a cyanobacterial patch. It was suggested that the patch facilitates the accumulation of the BBD consortium during optimal conditions, leading to a bacterial combination which included more *Oscillatoria* sp..

A notable study by Ainsworth et al., 2007 pointed out that not all cases of BBD disease express its characteristic black band, suggesting that previous cases of disease where the BBD microbial consortium was present may have been misidentified.

There are 27 genera of Scleractinia affected by BBD in the Indo-Pacific, as found by Sutherland et al., 2004, Montano et al., 2013, Page and Wilis, 2006, and Raymundo and Weil, 2016. This list includes: *Acanthastrea, Acropora, Alveopora, Astreopora, Coeloseris, Cyphastrea, Diploastrea, Dipsastraea, Echinopora, Echinophyllia, Favia, Favites, Fungia, Gardineroseris, Goniastrea, Goniopora, Hydnophora, Isopora, Leptoria,*

Diseases of the Indo-Pacific

5.3. Brown Band Disease - Code: BrBv

Brown Band Disease (BrB) is a common disease in the Indo-Pacific. It was first noted in the literature in the mid 1990's, with the first formal description appearing relatively recently in a study by Willis et al., 2004 from the Great Barrier Reef. The disease forms as a linear band and progresses rapidly, in some cases killing Acroporid colonies of over 1m in diameter in a matter of weeks (Pau Urgel Plaza, personal observation), see figure below.

The Brown Band Disease microbial agent consists of a boring ciliate from the Oligohymenophorea class (Bourne et al., 2008). Ciliates aggregate over the skeleton and provide a means of macroscopic identification, by forming brown 'speckled' bands which progress through and kill the coral host. While the ciliates generally provide easy means of identification, it should be kept in mind that the recognizable nature of the band is variable depending on the density of the ciliates. When aggregations are dense, the band is discrete, while more dispersed ciliates may form a diffuse band margin. Often an additional thin bleached white band (stress response to the disease) is visible between the brown band and healthy coral tissue. It is questioned whether this band is a stress response by the coral due to the presence of a nearby BrB band, or if there is some other cause of coral stress, which provides an opportunity for microbial communities to settle.

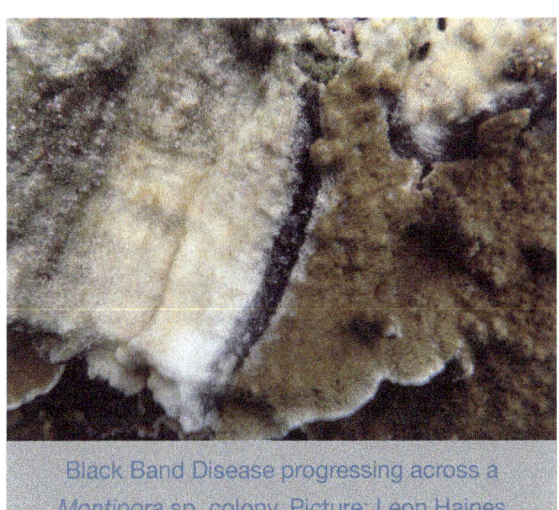

Black Band Disease progressing across a *Montipora* sp. colony. Picture: Leon Haines

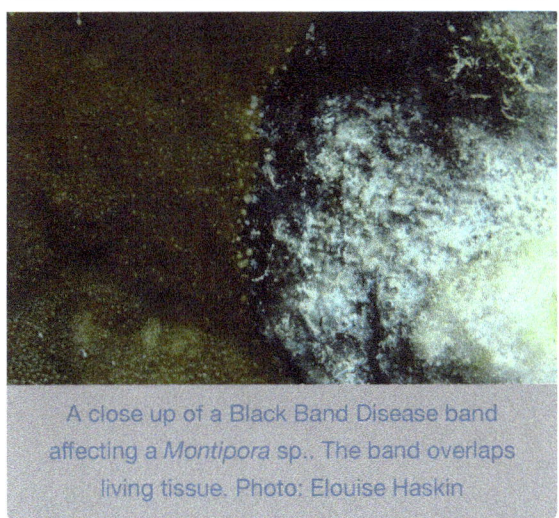

A close up of a Black Band Disease band affecting a *Montipora* sp.. The band overlaps living tissue. Photo: Elouise Haskin

Lobophyllia, *Montipora*, *Pachyseris*, *Pavona*, *Platygyra*, *Pocillopora*, *Porites*, *Psammocora*, *Seriatopora*, *Stylophora* and *Turbinaria*. Additionally, two soft corals; *Lobophytum* and *Sinularia*, and one hydrocoral; *Millepora* are hosts to BBD.

Diseases of the Indo-Pacific

An interesting study by Ulstrup et al., demonstrated that the ciliates utilize the photosynthetic properties of the zooxanthellae consumed from the tissue, expressing higher rates of photosynthesis than healthy adjacent coral tissue. It is yet to be confirmed if the ciliates maintain the photosynthesis deliberately or if they incidentally stumble into the position before the zooxanthellae are metabolised. This association with zooxanthellae accounts for the brown colouring of Brown Band Disease.

Abiotic controls are not thoroughly understood in the case of Brown Band Disease. It is suggested that physical stressors such as SCUBA diving activity, storm surge and anchor use can encourage BrB potentiality, by introducing coral lesions and therefor a means of entry for ciliate communities. Additionally, corallivorous animals act as vectors for disease spread, as has been shown in cases of Drupella sp. outbreaks. Crown of Thorns Sea Star lesions are also believed to encourage the expression of BrB. Brown Band Disease spreads more rapidly amongst densely packed corals.

A series of photographs demonstrating the rate of Brown Band Disease Spread on an *Acropora* sp. colony. Picture: Pau Urgel Plaza

A close up of an Acropora sp. colony suffering from BrB. In between the healthy tissue and skeleton a band of ciliates can be seen. Picture: Elouise Haskin

Diseases of the Indo-Pacific

There are 12 species of *Acropora* affected by BrB, and a further four genera, according to a study by Willis et al. (2004), the other genera include: *Isopora, Montipora, Pocillopora, Echinopora*. Additionally, more unidentified pocilloporids and faviids are susceptible.

5.4 Skeletal Eroding Band
Code: SEB

Skeletal Eroding Band (SEB) is very common in the Indo-Pacific, and was the first coral disease to be found and described from there in the late 1980's. It is believed to be an important contributor to coral mortality as it has the widest host range of scleractinian corals out of any known coral disease and has been shown to affect larger portions of coral assemblages than many other diseases surveyed. For example, a study conducted on the Great Barrier Reef by Page and Willis 2008, over a three year period found that a

SEB affecting a *Danafungia spp*. SEB often progresses around the disk in a 'pizza slice' pattern. Picture: Elouise Haskin

on a 500km reef stretch, SEB was present on 90-100% of surveys, and overall affects 2% of scleractinian colonies surveys.

SEB is one of the few diseases to have a eukaryotic microbial agent- the ciliate species *Halofolliculina corallasia*. Ranging in colour from dark green to black, it's ciliates 'speckle' the surface of the coral skeleton, forming

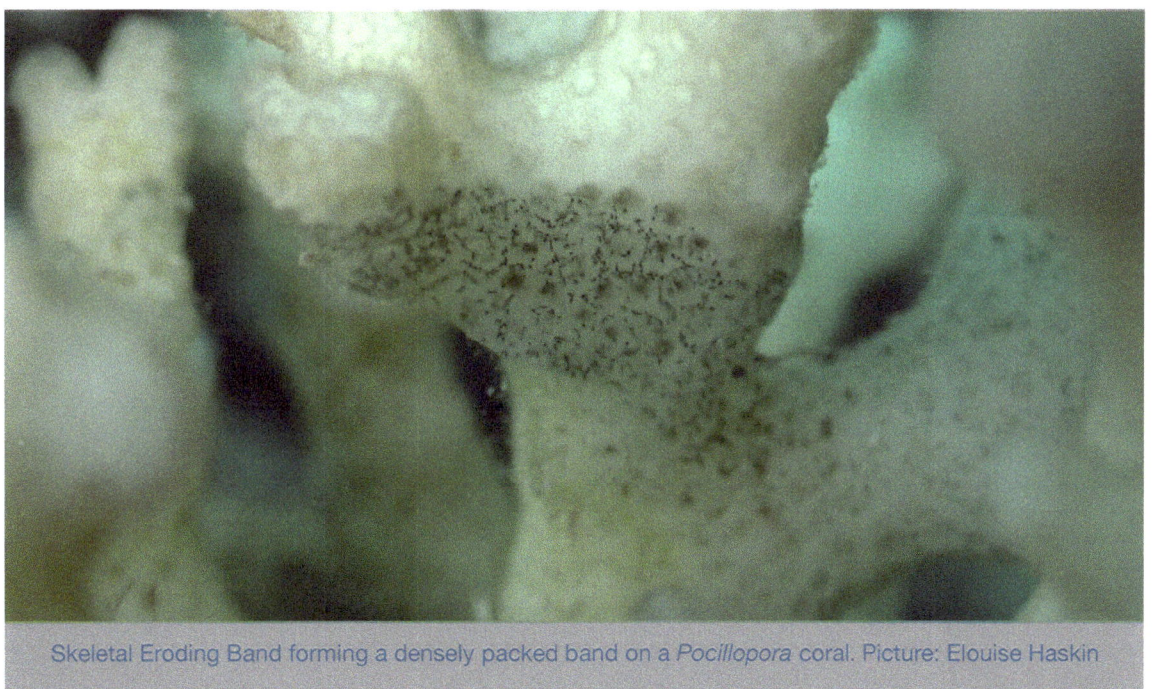

Skeletal Eroding Band forming a densely packed band on a *Pocillopora* coral. Picture: Elouise Haskin

5 Diseases of the Indo-Pacific

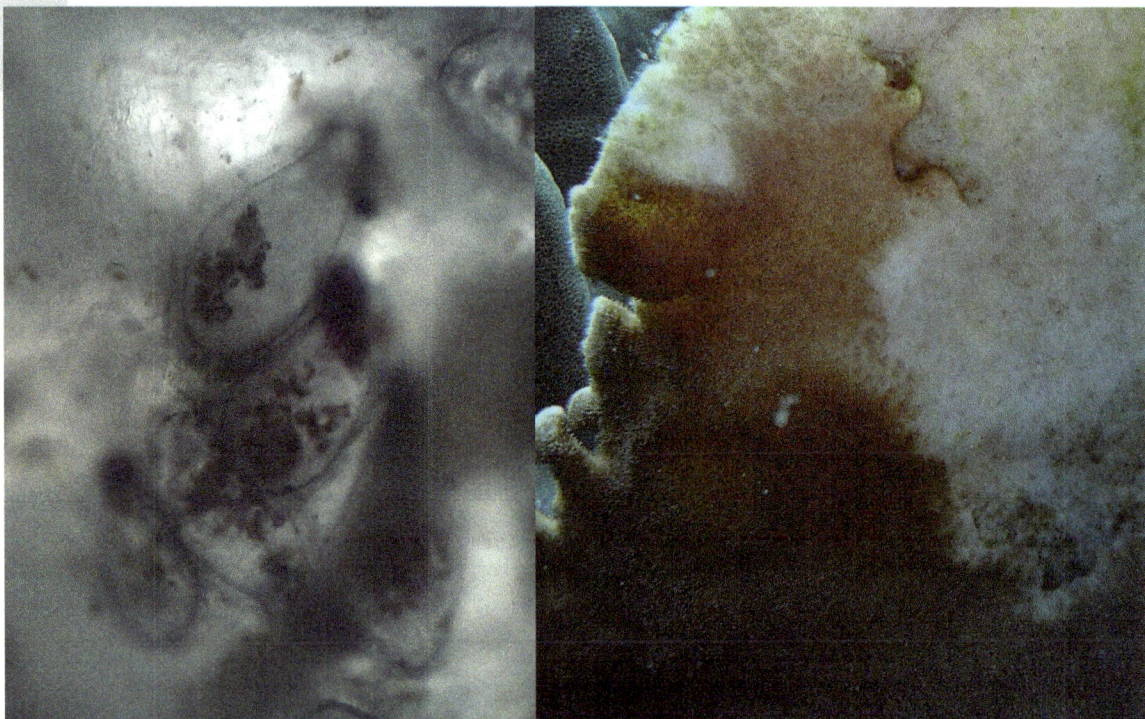

Left: A microscopic view of SEB ciliates, zooxanthellae are visible. Right: SEB on a *Pavona* sp. A patch of the ciliated band can be seen in the lower middle area. Pictures: Elouise Haskin

linear 1-10cm wide bands that progress across coral colonies, killing the tissue. Under microscopic analysis, zooxanthellae assemblages are easily identifiable. Similarly to Brown Band Disease, the zooxanthellae of SEB are responsible for the dark colour of the SEB ciliates. Depending on the density of ciliates, bands can be dark and discrete or diffuse and require close observation to correctly identify. As an example, *Pavona* is a genus where SEB is frequently difficult to spot, as ciliate communities are dispersed spaciously across polyps. In examples of thick ciliate communities, be cautious not to confuse the band with Black Band Disease. Progression rates of SEB are subacute, and lie at the low-lying end of the coral disease spectrum,

In terms of susceptibility, corals with complex growth forms are at higher risk of contracting Skeletal Eroding Band Disease. Acroporids and Pocilloporids are two groups which are particularly susceptible, they have large surface areas and thinner skeletons than massive colonies. Lesions caused by debris such as a plastic and nets can lead to SEB. Both provide entry points for ciliate infections. Confirmed environmental influences, and disease triggers are hypothesised and are broad. Similarly, to the ciliate entry point discussed above, other disturbances which cause stress to corals may lower their immune responses. Bleaching inducing warm temperatures, or water pollutants, or sedimentation damage and necrosis are all factors which may reduce immunity, and therefor increase susceptibility to infection.

Diseases of the Indo-Pacific

There are over 36 genera of coral which are recorded hosts of SEB, according to Sutherland et al., 2004, Page et al., 2016 and include: *Acropora, Alveopora, Astreopora, Coscinataea, Cyphastrea, Diploastrea, Dipsastraea, Echinopora, Favites, Fungia, Galaxea, Goniastrea, Goniopora, Hydnophora, Isopora, Leptastrea, Leptoria, Leptoseris, Lobophyllia, Montipora, Mycedium, Oulophyllia, Oxypora, Pachyseris, Pavona, Pectinia, Platygyra, Plerogyra, Pocillopora, Podabacia, Porites, Psammocora, Scolymia, Seriatopora, Stylophora* and *Turbinaria*. Additionally, one species of *Millepora* is confirmed to host SEB (Page and Willis 2008).

5.5. Pacific Yellow Band Disease Code: PYBD

Yellow diseases were first discovered in the Caribbean in the 1970's, with Yellow Band Disease termed in the 1990's. Much like White Syndrome, Yellow Band Disease has been a source of confusion, with different symptoms being placed under the same name. For simplicity, as described in Diseases of Corals 2016, Pacific Yellow Band Disease has been coined as the most up to date and relevant name for the Indo-Pacific, and will be maintained during this manual. Accordingly, two other YBD have been formulated, the Caribbean Yellow Band Disease (refer to chapter 7) and Arabian Yellow Band Disease.

PYBD contains a four species consortium of *Vibrio* bacteria, including *Vibrio rotiferianus, Vibrio harveyi, Vibrio alginolyticus* and *Vibrio proteolyticus*. The disease characterized by a pale yellow to white blotch that slowly radiates outwards. It is focal or multi focal and the margin is somewhat diffuse. The Caribbean has experienced plague-like proportions of CYBD, affecting many additional genera with similar symptoms to how PYBD affects *Diploastrea heliopora* in the Indo-Pacific. However, etiological links have not been confirmed between the geographically isolated areas. PYBD is noted to also

A *Herpolitha limax* showing signs of PYBD. Picture: Elouise Haskin

Diseases of the Indo-Pacific

A *Diplosatrea heliopora* expressing PYBD, and patches of mortality. Picture: Chad Scott

have a band of 'stress bleaching' in between the yellow/pale band and healthy coral tissue. It is believed that the disease reduces cell division of the symbiotic zooxanthellae, contributing to the demise of the coral tissue. The abiotic control of PYBD includes increased ocean sea surface temperatures.

On Fungiids, YBD is known to make pale blotch patterns. When tissue degradation is near-completed, a fluorescent pink from under the tissue residing in the skeleton replaces the pale colour, before tissue mortality occurs (Cervino et al. 2008). Confirmed coral hosts of PYBD in the Indo-Pacific only exist in the coral triangle and includes three genera: *Diploastrea*, *Herpolitha* and *Fungia*.

5.6. Atramentous Necrosis - Code: AtN

Sometimes referred to as 'The Black Death', this disease is a complex one. Post-mortality, Atramentous Necrosis (AtN) accumulates a series of dark to black fouling community which forms a thick layer under a white film that is focal or multifocal, resulting in a grey area of mortality. The top layer is easily removed by strong currents or the wave of a hand. Lesions appear as focal, multifocal or irregular in shape and have discrete boundaries. This disease progresses rapidly

Diseases of the Indo-Pacific

Atramentous Necrosis affecting a *Podabacia* sp. coral. In this image the top layer has been removed to see the black community below. Picture: Elouise Haskin

and often needs tracking over a 24-48hr period to confirm the symptoms have been correctly identified. However chronic infections occur when fouling organisms settle over the affected area, halting progression and making correct identifications of the mortality cause difficult.

Atramentous Necrosis has been found to have four stages of development, according to a study published by Anthony et al. (2008). The first stage is a small area of bleached tissue, up to a few cm across in diameter. This is followed by this area losing its living tissue, exposing the skeleton. Stage three consists of a white film made up of bacterial filaments covering the recently exposed skeleton. The last step involves a dark fouling community building up underneath the white film. One or multiple patches can develop, and often join together before spreading across more of the colony. The abiotic controls of AtN include rainfall and associated runoff (Haapkylä et al., 2011), as well as temperature rise, which can aid pathogen transmission, survival and development and like mentioned previously, can increase host susceptibility by inducing stress in corals.

According to Raymundo et al., 2008 and Subhan et al., 2020, there are seven genera of coral which host AtN including; *Montipora*, *Acropora*, *Echinopora*, *Merulina*, *Turbinaria*, *Favites* and *Pectinia*.

5 Diseases of the Indo-Pacific

Chapter 5 Review

After completing the reading and discussion of the material covered in Chapter 6, you should understand and be able to answer the following questions. Please talk with your instructor about any questions you may have.

- What diseases are characterised by having a band. Explain their microbial communities.

- What is YBD easily confused with?

- Which three diseases have ciliate communities?

- Which genera of coral does BrB affect?

Chapter 6: Diseases of the Wider Caribbean Region

Coral disease has been researched more in the Wider Atlantic and Caribbean Regions* (hereby referred to as the Caribbean) more than any other area of the world. The Indo-Pacific has experienced relatively little devastation by coral diseases compared to Caribbean waters and in fact, the first coral disease reported to science, Black Band Disease, was discovered in the Caribbean in the 1970s. The Florida Keys are heavily impacted and intense ongoing research has resulted in many research labs to better understand the problems they face. Diseases are a continually evolving and incredibly complex science, with more being discovered as our understanding progresses. This chapter will cover some of the more common coral diseases of the Caribbean.

*During this chapter, we will refer to the Wider Caribbean Region as described by the **UN Environment Programme, 2020**; "The Wider Caribbean Region (WCR) comprises the insular and coastal States and Territories with coasts on the Caribbean Sea and Gulf of Mexico, as well as waters of the Atlantic Ocean adjacent to these States and Territories and includes 28 island and continental countries."

6.1. Aspergillosis - Code: ASP

Aspergillosis does not affect many Caribbean corals. In fact, while it plays a significant role in affecting Gorgonian sea fans, there are no known Scleractinia coral hosts. Like many coral diseases, it has been discovered in recent decades, first being described in 1995. It's microbial agent is a fungal species called *Aspergillus sydowii* which is frequently found in terrestrial soils, and is thought to enter marine environments through a variety of origins including water runoff (Kim and Rypien 2016). *A. sydowii* can cause illness in many animals and is a well-researched fungal pathogen.

Signs of Aspergillosis include a deep-dark purple colour surrounding tissues suffering from necrosis. The dead skeleton is quickly colonised by fouling organisms, giving the lesion a dark and distinct appearance. Additionally, parts of the fans may completely disintegrate, and it should be noted that in some cases, Aspergillosis can express itself as discoloured tumor-like growths (Kim and Rypien, 2016), much like the growth anomalies described in chapter 4. Progression is slow and identification is often easy as there are no similar diseases of the sea fans. It should be noted that this disease can occur in plague proportions, or induce partial mortality of affected corals (Kim and Rypien 2016).

According to Nagelkerken and Smith (1997), two species of Gorgonians host Aspergillosis; *Gorgonia ventalina* and *G. flabellum*, with a third species; *G. mariae* thought to be susceptible to the fungus, to a lesser extent.

Diseases of the Wider Caribbean Region

6.2. Black Band Disease - Code: CBBD

Black Band Disease (CBBD) of the Caribbean has the same characteristics as the BBD of the Indo-Pacific and was the first coral disease described to science. CBBD is characterised by a distinct dark red to black filamentous band and is easily confused with ciliated diseases such as Caribbean Ciliate Infection (discussed in the next segment). CBBD of the Caribbean is suggested to have a different species of cyanobacteria to the Indo-Pacific variety, which is thought to be closely related to *Oscillatoria* (Sutherland et al. 2004).

CBBD is linear in shape and it progresses rapidly. Often an area of white will trail behind the disease line, as fouling organisms typically won't colonise the exposed skeleton as fast as the disease progresses. The band will vary from a few millimetres to a few centimetres in width.

Diploria sp. succumbing to Caribbean Black Band Disease. Picture: Kirsty Magson

According to Sutherland et al., 2004, there are 12 genera and 19 species of scleractinian corals which host BBD, including *Acropora*, *Colophyllia*, *Dichocoenia*, *Diploria*, *Favia*, *Madracis*, *Meandrina*, *Montastraea*, *Porites*, *Siderastrea*, *Solenastrea* and *Stephanocoenia*. Additionally, Sutherland identifies 6 species and three genera of gorgonians which host BBD, including *Gorgonia*, *Plexaura* and *Pseudopterogorgia*

NOTE: Red Band Disease

The difference between CBBD and Red Band Disease (RBD) has been historically debated due to similar cyanobacterial community. More recently, RBD has been synonymized with Black Band Disease, according to Richardson et al. (2016) as it has been confirmed that microbial community is the same. Confusion can arise when corals affected by CBBD are exposed to more light, as the cyanobacterial community transitions to a lighter red colour (Richardson et al. 2016).

6.3. Caribbean Ciliate Infection Code: CCI

Caribbean Ciliate Infection, hereby referred to as CCI is believed to be either a recent discovery to the Caribbean or has been historically overlooked. CCI is characterised by a series of densely packed dark ciliate community, similar to the Indo-Pacific SEB, which are easily visible and can either overlap with living tissue or trail behind exposed skeleton (Verde et al. 2016). The dark colouration of the band of CCI is visually similar to CBBD when the ciliate community is organised densely. It was

Diseases of the Wider Caribbean Region

described as a disease of Caribbean corals in 2006, in a study by Cróquer et al.,. The responsible microbial agent has been identified as a species of *Halofolliculina*, which have previously been thought to only effect corals in the Indo-Pacific in the form of SEB (Cróquer et al., 2006b).

Halofolliculina are suggested to be infectious, and this hypothesis has been tested and confirmed under laboratory conditions by Cróquer et al., (2006b). Additionally, coral hosts with lesions are more susceptible to infection than those which have no lesions. It has been suggested that chemical stress signals released by the colony may attract the ciliate community, similarly to how coral predators are drawn to stressed colonies. Additionally, the same study by Cróquer et al., (2006b) proposes that warmer temperatures increase CCI prevalence in corals, but only to those which are already experiencing stress.

According to Cróquer and Bastidas (2009), progression averages around 1cm per month. The lesion has a distinct linear or multifocal pattern, characterised by the ciliate community.

According to Cróquer et al., 2006 , CCI affects 15 genera and 25 species, including; *Acropora cervicornis, A. palmata, A. prolifera, Agaricia agaricites, A. fragilis, A. tenuifolia, A. lamarcki, Colpophyllia natans, Dichochoenia stockesi, Eusmilia fastigiata, Leptoseris cuculata, Madracis decactis, M. mirabilis, Orbicella annularis, O. faveolata, O. franksi, Montastraea cavernosa, Diploria strigosa, D. labyrinthiformis, Porites furcata, Porites astreoides, Siderastrea siderea, Scolymia cubensis, Stephanocoenia intersepta* and hydrocoral *Millepora alcicornis*.

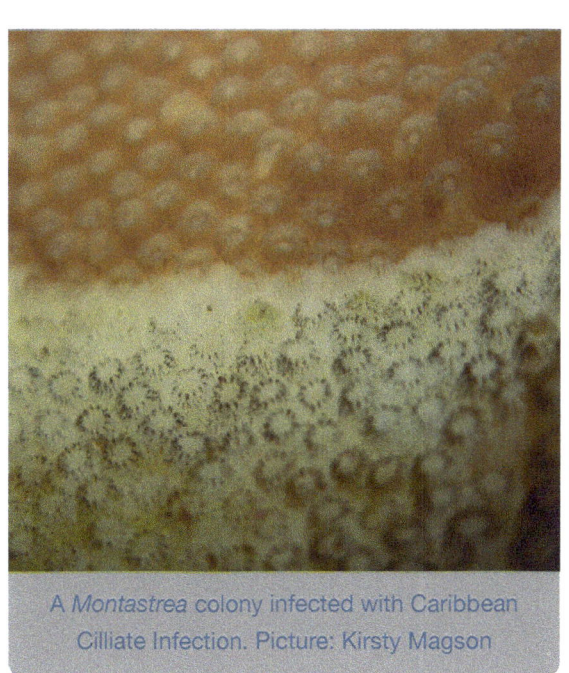

A *Montastrea* colony infected with Caribbean Cilliate Infection. Picture: Kirsty Magson

An *Agaricia* colony close up, infected with CCI. Cilliates can be seen a few centimetres before the healthy tissue. Picture: Kirsty Magson

Diseases of the Wider Caribbean Region

6.4. Caribbean Yellow Band Disease/ Yellow Blotch Disease - Code: CYBD

Yellow Band Disease (YBL) is another significant contributor to hard coral loss in the Caribbean. It affects a variety of coral species, with *Montastraea* spp. being a key study genus due to their high susceptibility. YBL was first noted to science in surveys conducted in 1996 in Panama, affecting *M. faveolata* and *M. annularis* (Santavy et al., 1999).

This disease progresses slowly to moderately and is characterised by a small area of pale to yellow tissue which can be slightly swollen in some cases. Lesions are annular to irregular and progress at a rate of approximately 0.5-1cm per month (Donà et al., 2008). Eventually, the areas towards the middle of the lesion lose their tissue and are colonised by filamentous algae and fouling organisms. Once again, *Vibrio* bacteria plays a role as the causal agent responsible for this tissue loss and is believed to directly affect the symbiotic zooxanthellae within the coral tissue, ultimately reducing the production of most of the coral hosts energy source. A study by Weil et al., (2009), tested the fecundity in tissue patches varying from infected to healthy, and compared them to control groups; they found that the count of eggs within the infected tissues was up to 99% lower than in healthy tissues.

Caribbean Yellow Band Disease. Public Domain image by Andy Bruckner, NOAA.

Diseases of the Wider Caribbean Region

According to Sutherland et al., 2004, YBL affects six genera and 9 species of stony corals in the Caribbean, including *Agaricia*, *Colpophyllia*, *Diploria*, *Favia*, *Montastraea* and *Porites*.

6.5. Dark Spots Disease Code: DSD

The causal agent of dark spot disease is yet to be isolated and confirmed, but so far, a consortium of bacterial, viral and cyanobacterial pathogens have been suggested. Interestingly, it has been observed that there is no difference between pathogen assemblages of healthy and infected coral hosts. This suggests that the cause may not be bacterial or viral, but that the spots are more of a response to stress, such as physical lesions, as dark spots disease does not always cause complete mortality (Randall et al. 2016).

Dark spot disease is expressed as a series of multifocal dark spots of the tissue, which do not always result in tissue loss, however it is thought that the disease causes a shutdown of mitosis within the tissue cells which ultimately reduces growth and reproductive ability of corals (Randall et al., 2016).

According to Randall et al. (2016), dark spot disease mostly effects *Siderastrea siderea*, *O. faveolata*, and *Stephanocoenia*

Dark Spots Disease. Picture: Kirsty Magson

6 Diseases of the Wider Caribbean Region

Caribbean White Band Disease. Photo used under license: CC BY-SA 3.0, https://en.wikipedia.org/w/index.php?curid=32676916

intersepta, and according to Sutherland et al., (2004), there are a total of 11 species affected in the Caribbean, including the following eight genera; *Agaricia, Colpophyllia, Diploria, Isophyllastrea, Meandrina, Montastraea, Siderastrea* and *Stephanocoenia.*

Caribbean White Diseases:

In the Caribbean, white diseases have been more thoroughly studied than in the Indo-Pacific. 'White Syndromes' is sometimes used to refer to a collection of white Caribbean diseases, while in the Indo-Pacific it is still under scrutiny as to whether multiple diseases with similar expression have been given the same name, which still lack etiological understandings.

6.6. White Band Disease Type I and Type II - Code: WBD

First appearing in the Caribbean in the 1970's and followed by the *Diadema antillarum* urchin disease outbreak the next decade, the outbreak of White Band Disease (hereby referred to as WBD) reduced coral coverage by 80% by the 1990's, drastically affecting their historically famous staghorn coral seascapes (Cramer et al. 2020). The region had been dominated by three species of densely populated *Acropora*, providing a means for the disease to travel through the population quickly and effectively. After the mass mortality, the natural process of recruitment onto the remaining skeleton substrate was halted as the second major disease outbreak of the region and decade led to the die-off of an important herbivore, the urchin. This enabled highly competitive macro-algae to out-compete and overgrow the corals.

The decade of White Band Disease and the *Diadema* die-off where the events which brought coral diseases to the eyes of the global scientific community. The Caribbean thereafter became a hub for research. Since then, a magnitude of diseases and related studies have developed from this region, as coral diseases continue to brutally plague the region.

WBD is characterised by a band of white skeleton which separates healthy colony tissue from exposed skeleton, which may be colonised by fouling organisms. This band varies from a few mm to a few cm's in width and sloughs off at the margin of disease progression. In 1998, another form of this disease was discovered and thus, WBD was divided into two types: WBD type I and WBD type II (Ritchie 1998). The difference between type I and II, is that WBD type II is characterised by an additional band of bleached tissue which precedes the mortality band (Gil-Agudelo et al., 2006). WBD type I on the other hand has the mortality line directly adjacent to the exposed skeleton without any bleaching preceding the line. The microbial community which inhabits these diseases has been tricky to confirm. For WBD type I there has been little consensus to what pathogens are at play, although some suggestions have been made. Type II on the other hand has been identified as Vibrio charchari being the causal agent (Kline and Vollmer, 2011).

According to Sutherland et al., 2004, there is one genus and two species of coral which host WBD type I; *Acropora cervicornis* and *Acropora palmata*, and one species which hosts WBD type II; *Acropora cervicornis*.

6.7. White Patch Disease
Code: WPD

Another white disease of the Caribbean, White Patch Disease (WPD) also known as white pox disease, is a significant contributor to reduced hard coral coverage in the Caribbean. Patterson and Ritchie (2002) found that it was the first coral disease caused by a human-derived faecal bacterium, *Serratia marcescens*.

This disease expresses a series of lesions which appear simultaneously on *Acropora palmata*, the only known susceptible coral host, and progress quickly, removing coral tissue to expose skeleton at a rate of 2.5cm^2 of tissue loss on average per day. Lesions are irregularly shaped, and the nature of there being multiple lesion locations separates this white disease from the others, as the progression doesn't necessarily begin at the base of a colony and move its way upward, but can progress in all directions (Patterson et al., 2002).

Diseases of the Wider Caribbean Region

6.8. White Plague Disease
Code: WPI

Considered to be one of the most lethal diseases in the Caribbean, White Plague Disease (WP) was first noted in the 70's in outbreak proportions progressing at a rate of approximately 2-3mm per day, before being observed again in the mid 90's but as a faster progressing (1-2cm per day) and more ubiquitous disease, causing the division into white plague disease type I (WP-I) and white plague disease type II (WP-II) (Precht et al., 2016; Richardson, 2016).

White plague expresses itself similarly to Caribbean White Band Disease, however more species of coral are affected, and the rates of progression are generally a lot faster, especially in cases of WP-II (Richardson 2016). Similarly, to WBD, WP progresses from the base of a colony and then moves upwards and can have multiple lesions at one time.

Its rapid progression and linear to annular shape results in large exposed patches of white skeleton. The bacterial pathogen of WP-II is *Aurantimonas coralicida* (Sutherland et al., 2004; Richardson, 2016). *Thalassomonas loyana* and *Vibrio coralliilyticus* have also been suggested to be a part of the consortium which comprise the WPs, however this information is currently suggestive, and has been disputed (Voolstra 2014).

According to Sutherland et al., 2004, the white plague diseases (type I and II) affect 38 species of coral, including the following 23 genera: *Agaricia, Colpophyllia, Cladocora, Dendrogyra, Dichocoenia, Diploria, Eusmilia, Favia, Isophyllastrea, Isophyllia, Leptoseris, Madracis, Manicina, Meandrina, Montastraea, Mussa, Mycetophyllia, Oculina, Porites, Scolymia, Siderastrea, Solenastrea* and *Stephanocoenia*. Additionally, two species of *Millepora* have been recorded as susceptible to these diseases.

6.9. Stony Coral Tissue Loss Disease
Code: SCTLD

Stony Coral Tissue Loss Disease (SCTLD) hit the Caribbean hard, first being discovered in the Florida Keys in 2014. It quickly became one the region's most concerning diseases due to its fast rates of infection and large host range of approximately half of the species of corals in the Caribbean.

The discovery of this disease followed a dredging project which was implemented adjacent to the first recorded cases, which was followed by a warm thermal anomaly. These events lead to questions regarding what caused this new disease, to whether a new pathogen may have been introduced to the environment via the dredging project, or whether the combined changes in environmental conditions lead to a new level of susceptibility of corals within the region (Weil et al., 2019).

SCTLD is focal or multi-focal and generally begins as a series of patches along the bottom edges of a coral colony. It spreads rapidly across the colony either annularly or linearly, killing tissue and exposing white skeleton. Occasionally, such as for *Montastraea cavernosa*, there will be a band of pale tissue between the adjacent skeleton and tissue (NOAA, 2018). SCTLD can remove coral tissue at rates of up to 3-4cm per day and can be influenced by the environment (such as light availability and temperature) and the condition or species of the coral host (Weil et al., 2019).

Since 2014, the disease has spread north and south of its original discovery and, depending on environmental conditions, has been recorded spreading as fast as 10km and as slow as 2.5 km per month. It is believed the disease is water-borne as the pathogens responsible are likely able to travel with water currents. Additionally, the possibility of toxins or pollutants in the water increasing susceptibility has also been considered by Caribbean researchers (NOAA, 2018).

According to NOAA (2018), SCTLD affects around 32 species and 24 genera of corals, including *Acropora, Agaricia, Cladocora, Colpophyllia, Dendrogyra, Dichocoenia, Diploria, Eusmilia, Favia, Helioseris, Isophyllia, Madracis, Meandrina, Montastraea, Mussa, Mycetophyllia, Porites, Pseudodiploria, Oculina, Orbicella, Scolymia, Siderastrea, Solenastrea*, and *Stephanocoenia*.

A Pseudodiploria colony losing tissue due to SCTLD. Picture: Kirsty Magson

6 Diseases of the Wider Caribbean Region

Chapter 6 Review

After completing the reading and discussion of the material covered in Chapter 6, you should understand and be able to answer the following questions. Please talk with your instructor about any questions you may have.

- What environmental factors are believed to have potentially influence the outbreak of Stony Coral Tissue Loss Disease?

- What is the difference between White Band Disease Type I and Type II?

- Which two Caribbean diseases have been synonymized?

- Which Caribbean disease does not affect scleractinian corals?

Chapter 7: Data Collection

Identifying conditions of compromised coral health must be accompanied with the collection of comprehensive, long-term datasets. Collecting and analysing data is crucial to management programs. Data allows us to create a record for the reef, showing its baseline state, identify changes in health and disease perveance, track disease progression and transmission, and respond accordingly. Additionally, having accurate records helps reef managers understand if mitigation techniques are successful, and may contribute to creating community regulations to ensure future problems are avoided where possible.

A variety of survey techniques are available for collecting coral disease data:
- Belt-Transect Line
- Quadrants
- Colony Tagging
- Photographic Documentation

7.1. Belt-Transect

Often, we utilize our permanent Ecological Monitoring Program (EMP) lines, and will have a series of teams collecting Fish, Invertebrate, Substrate and Coral Disease data. Like with our standard EMP lines, we want our data to be collected from the exact same locations each time to ensure we can accurately record changes. While we use the same lines, we conduct disease surveys slightly different to the standard EMP methodology. Instead of collecting data from four 5m× 20m segments with five-meter gaps along a 100 meter transect line, we collect data from four 1m × 10m segments with gaps of 15 meters between. Additionally, rather than the Point Intercept Transect (PIT), data is collected in a 1-meter-wide belt. Data collection is intensive during this survey, this is why the belts are slightly shorter, as finishing dives with a realistic collection goal and complete dataset is important.

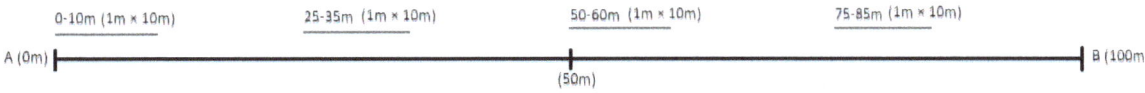

Figure 1: Grey lines signify the transect lengths where data is collected (birds eye view).

7 Data collection

Equipment includes:
- One 100m reel, or two 50m reels.
- Data collection slate (slate set up pictured below).
- Two Pencils (one spare).
- A one-meter-long PVC pipe, or a belt measuring alternative.
- A camera for photographing new or unidentifiable diseases (ideal but not compulsory).
- All standard SCUBA and safety equipment.

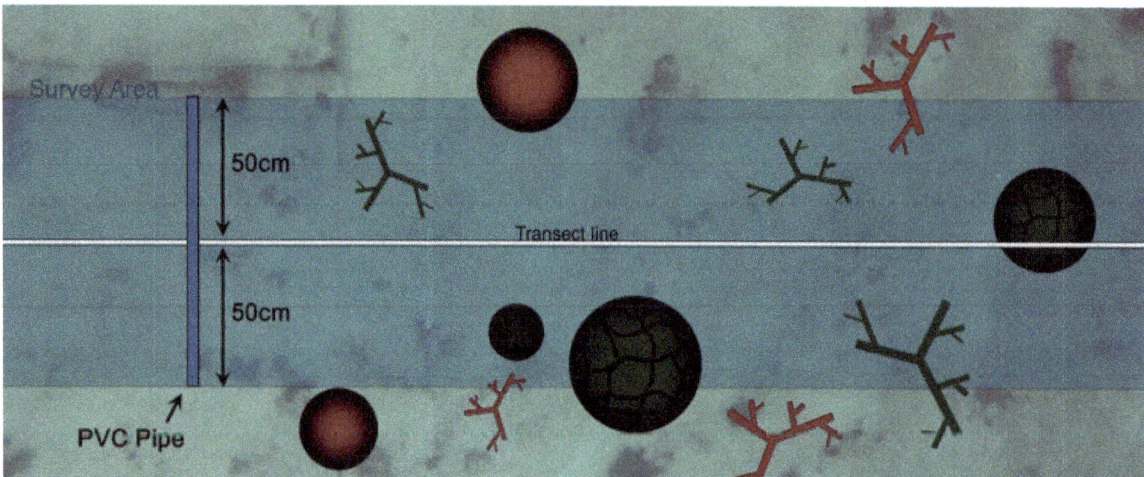

The one-meter pipe or marker is crucial, as it is used it to measure the horizontal belt over the line (50cm on each side) for each 10m section. Data is collected on all every coral colony that has half or more of its total diameter within the data collection belt. The only corals lying within that belt that we don't collect data on are fragments and recruits (with a diameter less than 3cm). We collect data by tallying on our slate.

Field Example

In the figure above, you will see that colonies which are more than 50% in the belt are colored green, these are the ones for which data will be collected. Those that are red are mostly outside the belt, and will not be monitored. In the event of very large corals, if the entire 1m belt covers the coral, then the coral is included.

Data collection 7

7.2. Survey Slates

The slate will come with all compromised health options provided (example pictured below).

How to conduct the survey. Each coral with 50% or more of colony is recorded. Picture: Pau Urgell Plaza

GENUS	OVER-GROWTH					PREDATION						DISEASE					OTHER						
	H	SP	TA	MA	CB	COT	DRUP	DAM	TRIG	PRT	BTFY	BBD	PYBD	BrB	SEB	WS	GA	PGA	TMD	SDS	PHY	UBL	OTH
Acro.																							
Pocill.																							
Porites																							
Pavona																							
Platy.																							

A belt-transect disease survey slate

7 Data collection

Corals are identified to a genera level, growth forms are not assessed, and noted along the left hand side column. Then the tally is made amongst the rows and columns where appropriate, an example completed slate is pictured below.

GENUS	OVER-GROWTH				PREDATION						DISEASE					OTHER						
	H	SP	TA	MA	CB	COT	DRUP	DAM	TRIG	PRT	BTFY	BBD	PYBD	BrB	SEB	WS	GA	PGA	TMD	SDS	PHY	UBL	OTH
Acro.	𝍤II					I	𝍤I		I						III								
Pocill.	𝍤II 𝍤													𝍤𝍤 III									
Porites	IIII			𝍤															IIIII				
Pavona		IIII		IIII											I								
Platy.																							

An example of a completed data collection slate. Coral genera are noted along the left hand side and also notes the growth form of each in brackets. Note: No coral genera are initially included in the far left column, to save space and to account for variations in coral genera presence at different survey sites, for this portion of the data collection all genera are manually entered.

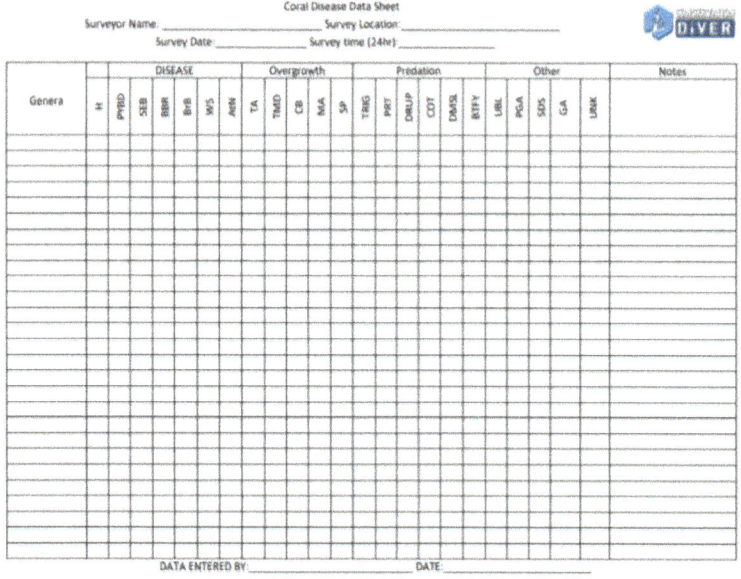

The hard copy data sheet. Note that no coral genera are included in the far left column, to save space and to account for variations in coral genera presence at different survey sites, for this portion of the data entry all genera must be manually entered during the survey.

74 Coral Disease Identification and Monitoring Manual

Data collection 7

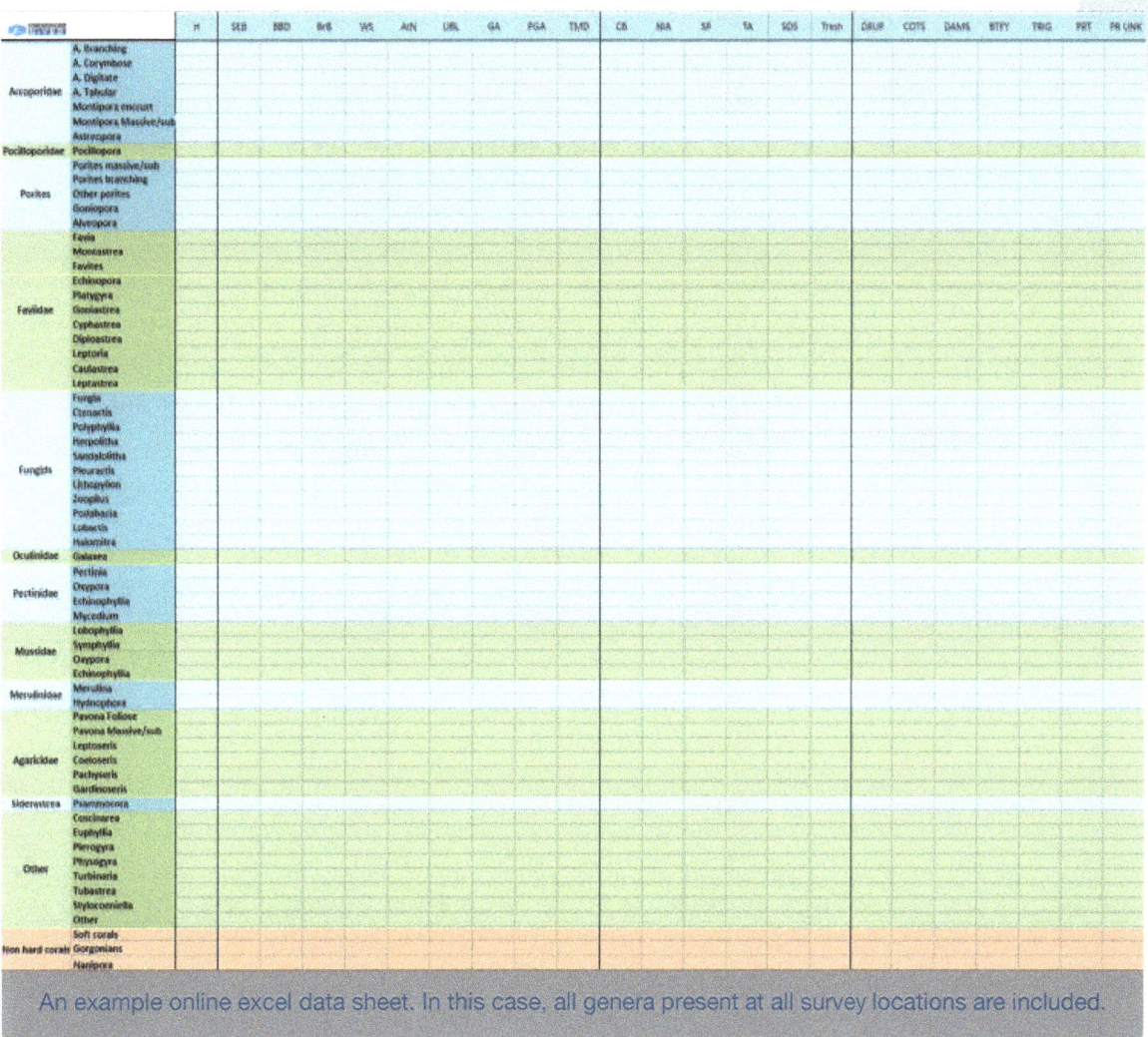

An example online excel data sheet. In this case, all genera present at all survey locations are included.

Post survey, data is transferred to a hard and soft copy (excel spread sheets) so it can be stored and used when needed. Hard copy sheets are displayed below. Health options are permanent, but genera and growth form are added manually. For the online copy, genera and health option are provided and it is simply the task of adding the values to the appropriate boxes.

7.3. Quadrat Survey

This type of survey can be used in areas that have undergone recent stress, where the goal is to conduct more in-depth or higher-resolution surveys. They are often used to evaluate small or cryptic species and smaller areas of reefs. They can be both permanent or random/haphazard, depending on the studies goals and requirements. Data is collected only from within the boundaries of the quadrat, which is often 1 meter square.

7 Data collection

Equipment for this survey includes:

- A square PVP pipe quadrat (can be divided into segments with string, depending on the requirements of the survey, size also variable depending on survey requirements).
- A slate.
- Two pencils (one spare)
- All relevant SCUBA and safety equipment.

Depending on the goals of the survey, the quadrat will be placed in different ways. If a random survey is being done, the quadrant is laid on the sea bed, and the corals under it are assessed. It can then be moved until the desired amount of data is collected. Alternatively, the quadrant can be used along the established transect lines, placing it under the transect at set intervals. Another useful way to establish quadrants is in areas where you wish to track disease progression. In this case, the area of coral disease outbreak is marked with 4 metal stakes at each corner, a rope is then run around and through the centre to create 4 equal sized square areas. All the corals within this area are assessed at each survey.

From the figure above, we can formulate percentage coverage of each square in the quadrat to find the average coverage of Goniopora coral on a reef, here the average is 21%. This can be replicated again to find more precise results.

7.4. Tracking Colonies

This method involves placing a semi-permanent marker on the coral colony to ensure that surveyors are able to find it again on subsequent visits. Colonies are marked with a nail, or another appropriate tagging material. A timeline is created for data collection, and surveyors return to monitor coral mortality on a regular basis. Tagging of colonies can be used to track the progression of mortality in a single colony, as well as track mortality of individual colonies in a population.

For disease progression in a single colony, a nail or other marker is put at the lesion border. The day when the nail is inserted is day-0, and the distance from the disease band to the nail is measured on subsequent visits. It is important to keep in mind that the lesion will likely progress in all directions from the lesion centre, so, it is critically important to ensure the measurement is taken in the exact same direction during each measurement to ensure accurate data collection. Multiple measurements should be taken in this manner to formulate averages.

7 Data collection

When looking at mortality rates for colonies in a population, the colony would be marked, this is often done by placing a marker next to the colony, in the sand or on a dead bit of reef, as not to cause any further stress to the colonies being observed. This would be done to multiple colonies in a population to assess overall rates or mortality over time.

Both of these methods often incorporate **photographic documentation**, which is a great way of being able to overlook data collection at later dates, and creating learning resources for future coral conservationists.

These methods require:

- A slate.
- A pencil.
- A ruler to measure progression rates.
- An underwater camera to monitor change/progression.
- A 'marker' of some description, such as a nail- to be hammered into a coral in an easy to find location, or a bright marker to find the location again.

7.5. Data use

Data collected through Disease Surveys can be utilized in a variety of ways. Maintaining a continuous database, even if not used immediately means that it can be referred to when trying to identify disease outbreaks. It also can be kept for future researchers, and if questions relating to compromised states of coral health are posed at later dates, having an available dataset to refer to provides a means of finding the answers to those questions much quicker.

Chapter 7 Review

After completing the reading and discussion of the material covered in Chapter 7, you should understand and be able to answer the following questions. Please talk with your instructor about any questions you may have.

- How many segments do we collect data from during a survey? How many square meters does this comprise?
- How many copies of our data do we keep, how are these copies stored?
- Why is it important to have a thorough understanding of coral taxonomy and disease before our survey dives?

Chapter 8: Coding Accurately

Addressing the tricky business of having multiple stressors affecting a single coral colony.

During surveys we select one state of health per coral colony observed to ensure future datasets are not skewed. Choosing a dominant stress on a single coral colony can be a difficult decision that will present itself often during data collection and exploratory dives. It is not unusual for one stress to attract additional pressures to a coral, and so; two diseases, or disease and predation, or physical disturbance and predation (and many other combinations of these) may occur on a singly colony at one given time. Remember from the predator chapter that stressed corals release chemical signals that attract coralivores. Likewise, many diseases spread during thermal bleaching events. This means that multiple stress factors affecting a single coral colony are not unlikely.

Furthermore, some viruses will remain dormant in a colony, only expressing when the coral becomes immunosuppressed or compromised. This is why precursors for disease outbreaks are often factors that alter the environment, and add pressure to coral such as temperature increases, eutrophication or overfishing. This example mirrors how some human viruses interact amongst us. For example, many people are infected with HSV-1 (herpes simplex type 1

An *Acropora* sp. coral colony affected by Brown Band Disease, with Macro Algae growing on the recently exposed skeleton. Photograph: Elouise Haskin

8 Coding Accurately

virus), and occasionally express cold sores. Some might experience cold sores during childhood and never display symptoms again. However, during periods of stress, such as work pressure or illness, we might return to a symptomatic state. These examples demonstrate the remarkable likeness that colonial marine animals like coral, share with us.

So, how does a surveyor decide which stress is primary and what do they record during data collection? One example is pictured above, of a colony which is affected by Brown Band Disease and Macroalgae. It's best to look at the whole colony and consider which threat seems to be of the biggest concern. If the colony was suffering from small amounts of algae overgrowth, and there were multiple branches being killed by the BrB, it'd probably be best to say this was a disease problem. If on the other hand, you see the Macroalgae settling on a dead patch of coral, yet the colony as a whole was suffering from a severe bleaching event and had completely expelled all of its symbiotic Zooxanthellae; it'd be quite safe to say that the thermal stress/bleaching was the prominent problem here, and we'd want to ensure that we noted it on the survey.

As we mentioned before, many of these diseases only persist during periods of evaluated sea surface temperature. In the case of mass coral bleaching, we would know from other datasets that the degree of bleaching for the reef. During disease specific surveys we would thus want to highlight the disease, rather than the degree of coral bleaching. Instead, we would make a note in the metadata of the survey to indicate that it was performed on a bleached reef.

In situations where there are two major threats factoring into the situation, assess both and chose one which you can reasonably justify choosing, and make a note of the additional stress observed. Even amongst experienced field researchers, tricky decisions arise during data collection. We must simply do the best with the tools and information we have available to us. It is always preferable for the researcher to choose one category while underwater, rather than writing down two or leaving it for later analysis. Once back on land it is nearly impossible to accurately choose between two codes, so when in doubt, choose one and take a photo if possible.

A *Porites* sp. colony being affected by a variety of boring invertebrates, expressing Pigmentation Response. Photograph: Elouise Haskin

Coding Accurately 8

Chapter 8 Review

After completing the reading and discussion of the material covered in Chapter 7, you should understand and be able to answer the following questions. Please talk with your instructor about any questions you may have.

- What should you do if more than one stress is affecting a coral colony?

- What pressures might cause a coral to express a disease?

- When should you list two separate stressors on a single colony?

Appendix A

All Survey Codes			
Healthy – H			
Predation	**Overgrowth**	**Diseases**	**Other**
COT	MA	WS	PGR
DRUP	CB	BBD	GA
DAMS	SP	BrB	SDS
BTFY	TA	SEB	TND
TRIG		PYBD	PHY
PRT		AtN	UBL
COR		ASP	
WENT		CBBD	
		CCI	
		CYBD	
		DSD	
		WBD	
		WPD	

Appendix B

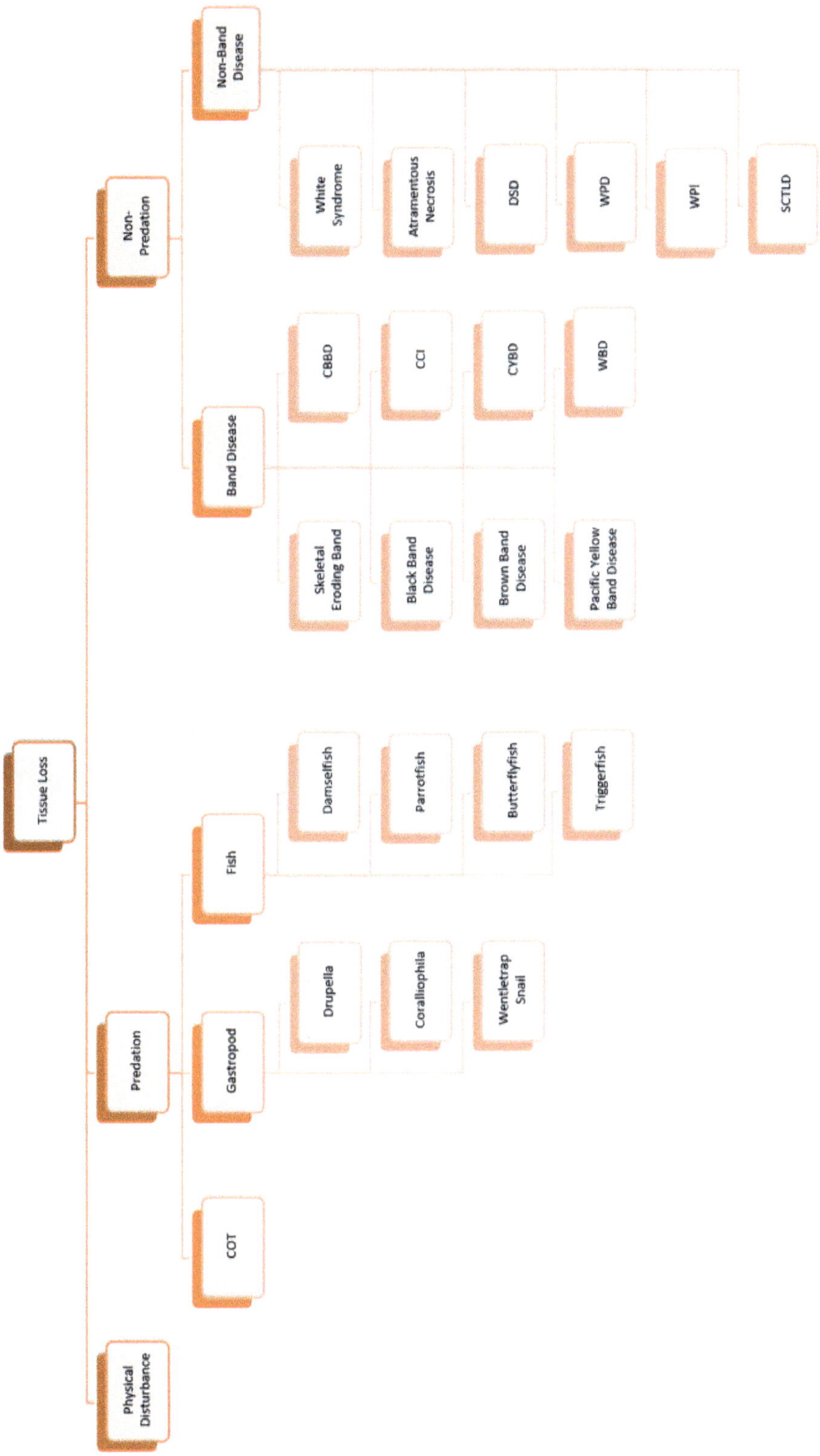

- Tissue Loss
  - Non-Predation
    - Non-Band Disease
      - White Syndrome
      - Atramentous Necrosis
      - DSD
      - WPD
      - WPl
      - SCTLD
    - Band Disease
      - CBBD
      - CCI
      - CYBD
      - WBD
      - Skeletal Eroding Band
      - Black Band Disease
      - Brown Band Disease
      - Pacific Yellow Band Disease
  - Predation
    - Fish
      - Damselfish
      - Parrotfish
      - Butterflyfish
      - Triggerfish
    - Gastropod
      - Drupella
      - Coralliophila
      - Wentletrap Snail
    - COT
  - Physical Disturbance

Coral Disease Identification and Monitoring Manual | 83

Appendix B

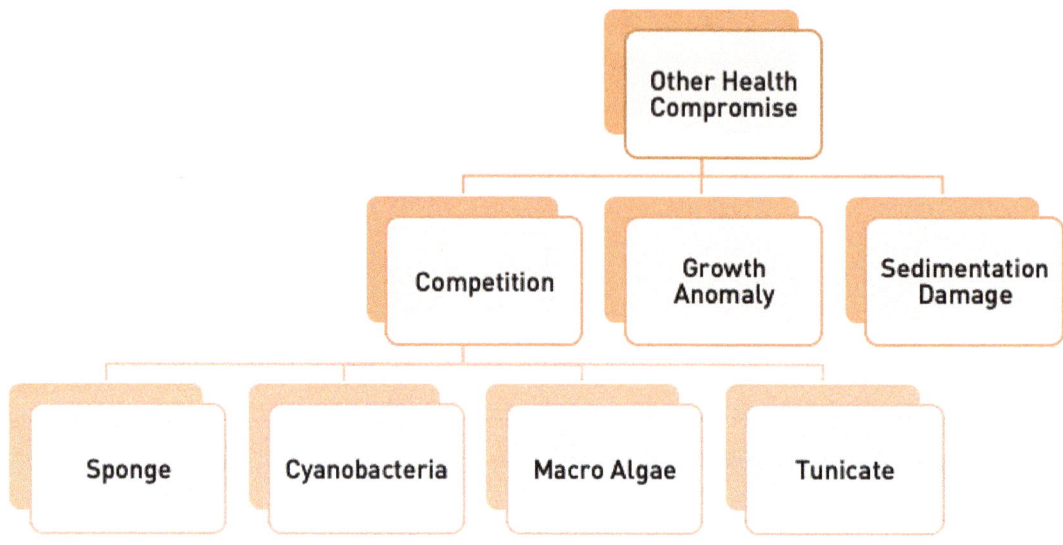

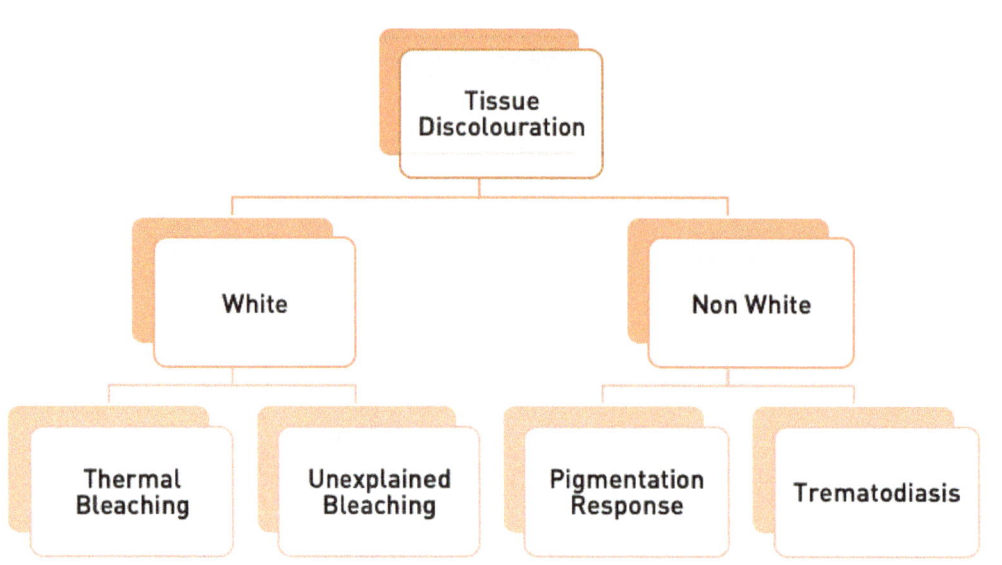

Glossary

Abiotic: Non-living matter (rock, sand etc. See *biotic*).
Biotic: Living, or previously living matter (algae, fish etc. See *abiotic*).
Coenasteum: The skeletal material in between corallites on corals.
Coralivores: An animal which eats coral polyps (COTS, Drupella, Butterfly fish, etc).
Detritus: Waste or debris, can be organic or inorganic.
Diffuse border: A gradual gradient line, or feathered border between a coral disease and healthy tissue.
Discrete border: A clear line between a coral disease and healthy tissue.
Excurrent Syphon: An anatomical structure in some invertebrate species that pushes out filtered water (see incurrent syphon).
Filamentous Algae: A single celled algae that form long visible structures.
Fouling Organism: An organism that lives on another organism in a non-parasitic relationship.
Incurrent Syphon: An anatomical structure in some invertebrate species that pull in sea water (see excurrent syphon).
Inorganic: Non-living matter (see organic).
Lesion: An opening or 'wound'
Oligotrophy: Water which is relatively low in nutrients and has abundant levels of oxygen.
Organic: Living, or coming from living matter (see inorganic).
Scleractinian: The order of Anthozoa containing the hard or stony corals.
Sessile: Non-moving, or attached organism (ie. barnacles or tunicates).
Stolon: A stem-like structure that joins some colonial organisms (zoanthids).
Symbiodinium: A genus containing the largest group of endosymbiotic dinoflagellates known.
Symbiosis: The interaction of two different organisms living in close physical association.
Urochordate: An invertebrate with notochords, also known as tunicates.
Zooxanthellae: A symbiotic dinoflagellate found in the tissue of many marine invertebrates (coral, anemone, giant clam).

References

Aeby, G.S., Williams, G.J., Franklin, EC, Haapkyla, J., Harvell, C.D., Neal, S., Page, C.A., Raymundo, L., Vargas-Ángel, B., Willis, B.L., Work, T.M. and Davy, S.K., 2011. Growth Anomalies on the Coral Genera Acropora and Porites Are Strongly Associated with Host Density and Human Population Size across the Indo-Pacific. PLoS ONE 6(2): e16887.

Aeby, G.S., 1992. The potential effect the ability of a coral intermediate host to regenerate has had on the evolution of its association with a marine parasite. In *Proc. 7th Int. Coral Reef Symp. Guam* (Vol. 2, pp. 809-815).

Anthony, S.L., Page, C.A., Bourne, D.G. and Willis, B.L., 2008. Newly characterized distinct phases of the coral disease 'atramentous necrosis' on the Great Barrier Reef. *Diseases of aquatic organisms*, *81*(3), pp.255-259.

Armstrong S.J., 2009. Ningaloo Marine Park Drupella long-term monitoring program: Data collected during the 2008 survey. Marine Science Program Data Report Series: MSPDR5. January 2009. Marine Science Program, Department of Environment and Conservation, Perth, Western Australia. 17p.

Bourne, D.G., Boyett, H.V., Henderson, M.E., Muirhead, A. and Willis, B.L., 2008. Identification of a ciliate (Oligohymenophorea: Scuticociliatia) associated with brown band disease on corals of the Great Barrier Reef. *Appl. Environ. Microbiol.*, *74*(3), pp.883-888.

Cervino, J.M., Thompson, F.L., Gomez-Gil, B., Lorence, E.A., Goreau, T.J., Hayes, R.L., Winiarski-Cervino, K.B., Smith, G.W., Hughen, K. and Bartels, E., 2008. The Vibrio core group induces yellow band disease in Caribbean and Indo-Pacific reef-building corals. *Journal of applied microbiology*, *105*(5), pp.1658-1671.

Cramer, K.L., Jackson, J.B.C., Donovan, M.K., Greenstein, B.J, Korpanty, C.A., Cook, G.M. and Pandolfi, J.M., 2020. Widespread loss of Caribbean acroporid corals was underway before coral bleaching and disease outbreaks. *Science Advances, 6*(17).

Cróquer, A., Bastidas, C., Lipscomp, D., Rodríguez-Martínez, R.E., Jordan-Dahlgren, E. and Guzman, H.M., 2006. First report of folliculinid ciliates affecting Caribbean scleractinian corals. *Coral Reefs, 25:* pp. 187-191.

References

Donà, A.R., Cervino, J.M., Goreau, T.J., Bartels, E., Hughen, K., Smith, G.W., and Donà, A., 2008. Coral Yellow Band Disease; current status in the Caribbean, and links to new Indo-Pacific outbreaks. *Proceedings of the 11th International Coral Reef Symposium*, Session 7.

Gil-Agudelo, D.L., Smith, G.W. and Weil, E., The white band disease type II pathogen in Puerto Rico. *Revista de Biologia Tropical, 54*(3), pp. 59-67.

Haapkylä, J., Unsworth, R.K., Flavell, M., Bourne, D.G., Schaffelke, B. and Willis, B.L., 2011. Seasonal rainfall and runoff promote coral disease on an inshore reef. *PloS one, 6*(2).

Kim, K. and Rypien, K. 2016. Aspergillosis of Caribbean Sea Fan Corals, Gorgonia spp. *Diseases of Coral,* pp. 236-241.

Kline D.I. and Vollmer S.V., 2011. White Band Disease (type I) ofEndangered Caribbean AcroporidCorals is Caused by Pathogenic Bacteria. *Scientific Reports*, 1:7.

Lamb, B.J., True, J.D., Piromvaragorn, S. and Willis, B.L., 2014. Scuba diving damage and intensity of tourist activities increases coral disease prevalence. *Biological Conservation. 178,* pp. 88-96.

Lindop, A.M.M., Hind, E.J. and Bythell, J.C., 2008. The unknowns in coral disease identification: an experiment to assess consensus of opinion amongst experts. In *Proceedings of the 11th International Coral Reef Symposium, Ft Lauderdale, Florida, 7–11 July 2008 Session number* (Vol. 7).

Montano, S., Strona, G., Seveso, D. and Galli, P., 2013. Prevalence, host range, and spatial distribution of black band disease in the Maldivian Archipelago. *Diseases of Aquatic Organisms, 105*(1), pp.65-74.

Nagelkerken, I., Buchan, K., Smith, G.W., Bonair, K., Bush, P., Garzón-Ferreira, J., Botero, L., Gayle, P., Harvell, C.D., Heberer, C., Kim, K., Petrovic, C., Pors, L. and Yoshioka, P., 1997a. Widespread disease in Caribbean sea fans: 11. Patterns of infection and tissue loss. *Marine Ecology Progress Series, 160,* pp. 255-263.

NOAA, 2018. Stony Coral Tissue Loss Disease Case Definition. Florida Keys National Marine Sanctuary.

Page, C. and Willis, B., 2006. Distribution, host range and large-scale spatial variability in

12 References

black band disease prevalence on the Great Barrier Reef, Australia. *Diseases of aquatic organisms*, *69*(1), pp.41-51.

Page, C.A. and Willis, B.L., 2008. Epidemiology of skeletal eroding band on the Great Barrier Reef and the role of injury in the initiation of this widespread coral disease. *Coral Reefs*, *27*(2), pp.257-272.

Palmer, C.V., Mydlarz, L.D. and Willis, B.L., 2008. Evidence of an inflammatory-like response in non-normally pigmented tissues of two scleractinian corals. *Proceedings of the Royal Society B: Biological Sciences*, *275*(1652), pp.2687-2693.

Paterson, K.L., Porter, J.W., Ritchie, K.B., Polson, S.W., Mueller, E., Peters, E.C., Santavy, D.L. and Smith, G.W., 2002. The etiology of white pox, a lethal disease of the Caribbean elkhorn coral, Acropora palmata. *Proceedings of the National Academy of Sciences of the United States of America*, *99*(13), 8725-8730.

Pollock F.J., Wada N., Torda G., Willis B.L. and Bourne D.G., 2017. White syndrome-affected corals have a distinct microbiome at disease lesion fronts. Appl Environ Microbiol 83: e02799-16.

Potkamp, G., Vermeij, M.J. and Hoeksema, B.W., 2017. Genetic and morphological variation in corallivorous snails (Coralliophila spp.) living on different host corals at Curaçao, southern Caribbean. *Contributions to Zoology*, *86*(2), pp.111-S9.

Randall, C.J., Jordán-Garza, A.G., Muller, E.M. and Woesik, R.V., 2016. Does Dark-Spot Syndrome Experimentally Transmit among Caribbean Corals? *PloS one*, *11*(1): e0147493.

Raymundo, L.J., Halford, A.R., Maypa, A.P. and Kerr, A.M., 2009. Functionally diverse reef-fish communities ameliorate coral disease. *Proceedings of the National Academy of Sciences*, pp.pnas-0900365106.

Raymundo, L.J., Rosell, K.B., Reboton, C.T. and Kaczmarsky, L., 2005. Coral diseases on Philippine reefs: genus Porites is a dominant host. *Diseases of Aquatic Organisms*. *64*, pp. 181-191.

Raymundo, L.J. and Weil, E., 2016. Indo-Pacific colored-band diseases of corals. *Diseases of Coral*, pp.333-344.

Raymundo, L.J., Work, T.M., Miller, R.L. and Lozada-Misa, P.L., 2016. Effects of Coralliophila violacea on tissue loss in the scleractinian corals Porites spp. depend on host response. *Diseases of aquatic organisms*, *119*(1), pp.75-83.

Richardson L.L., 2004. Black Band Disease. In: Rosenberg E., Loya Y. (eds) Coral Health and Disease. Springer, Berlin, Heidelberg

Richardson, L.L., 2016. Aurantimonas coralicida—White Plague Type II. *Diseases of Coral,* pp. 231-235.

Richardson, L.L., 2016. Cyanobacterial-Associated Colored-Band Diseases of the Atlantic/Caribbean. *Diseases of Coral,* pp. 345-353.

Ritchie K.B. and Smith G.W., 1998. Type II White-Band Disease. *Revista De Biologia Tropical, 46*(5), pp. 199-203.

Order, C., Arif, C., Bayer, T., Aranda, M., Daniels, C., Shibl, A., Chavanich, S. and Voolstra, C. R., 2014. Bacterial profiling of White Plague Disease in a comparative coral species framework. *The International Society for Microbial Ecology, 8*, pp. 31-39.

Santavy, D.L., Peters, E.C., Quirolo, C., Porter, J.W. and Bianchi, C.N., 1999. Yellow-blotch disease outbreak on reefs of the San Blas Islands, Panama. *Coral Reefs, 18,* pp. 97.

Sato, Y., Willis, B.L. and Bourne, D.G., 2010. Successional changes in bacterial communities during the development of black band disease on the reef coral, Montipora hispida. *The ISME journal, 4*(2), pp.203-214.

Subhan, B., Arafat, D., Rahmawati, F., Dasmasela, Y.H., Royhan, Q.M., Madduppa, H., Santoso, P. and Prabowo, B., 2020, January. Coral disease at Mansuar Island, Raja Ampat, Indonesia. In *IOP Conference Series: Earth and Environmental Science* (Vol. 429, No. 1, p. 012027). IOP Publishing.

Sutherland, K.P., Porter, J.W. and Torres, C., 2004. Disease and immunity in Caribbean and Indo-Pacific zooxanthellate corals. *Marine Ecology Progress Series. 266,* pp. 273-302.

Ulstrup, K.E., Kühl, M. and Bourne, D.G., 2007. Zooxanthellae harvested by ciliates associated with brown band syndrome of corals remain photosynthetically competent. *Applied and Environmental Microbiology, 73*(6), pp.1968-1975.

12 References

Verde, A., Bastidas, C. and Croquer. A., 2016. Tissue mortality by Caribbean ciliate infection and white band disease in three reef-building coral species. *PeerJ* 4:e2196.

Weil, E., Cróquer, A. and Urreiztieta, I., 2009. Yellow band disease compromises the reproductive output of the Caribbean reef-building coral Montastraea faveolata (Anthozoa, Scleractinia). *Diseases of Aquatic Organisms, 87,* pp. 45-55.

Weil, E., Hernández-Delgado, E.A., Gonzalez, M., Williams, S., Suleimán-Ramos, S., Figuerola, M. and Metz-Estrella, T., 2019. Spread of the new coral disease "SCTLD" into the Caribbean: implications for Puerto Rico. *The News Magazine of the International Coral Reef Society, 34*(1), pp. 38-43.

Willis, B.L., Page, C.A. and Dinsdale, E.A., 2004. Coral disease on the Great Barrier Reef. In *Coral health and disease* (pp. 69-104). Springer, Berlin, Heidelberg.

References

Our Partner's/Team's Work

Allchurch, A., Mehrotra, R., Carmody, H., Monchanin, C., & Scott, C. M. (2022). Competition and epibiosis by the sponge Pseudoceratina purpurea (Carter, 1880) on scleractinian corals at a tourism hotspot in the Gulf of Thailand. Regional Studies in Marine Science, 49, 102131.

Hein, M. Y., Lamb, J. B., Scott, C., & Willis, B. L. (2015). Assessing baseline levels of coral health in a newly established marine protected area in a global scuba diving hotspot. Marine Environmental Research, 103, 56-65.

Hoeksema, B. W., van der Schoot, R. J., Wels, D., Scott, C. M., & Harry, A. (2019). Filamentous turf algae on tube worms intensify damage in massive Porites corals. Ecology, 100(6).

Lamb, J. B., True, J. D., Piromvaragorn, S., & Willis, B. L. (2014). Scuba diving damage and intensity of tourist activities increases coral disease prevalence. Biological Conservation, 178, 88-96.

Moerland, M. S., Scott, C. M., & Hoeksema, B. W. (2016). Prey selection of corallivorous muricids at Koh Tao (Gulf of Thailand) four years after a major coral bleaching event. Contributions to Zoology, 85(3), 291-309.

Roder, C., Arif, C., Bayer, T., Aranda, M., Daniels, C., Shibl, A., ... & Voolstra, C. R. (2014). Bacterial profiling of White Plague Disease in a comparative coral species framework. The ISME journal, 8(1), 31-39.

Scott, C. M., et al. "Population dynamics of corallivores (Drupella and Acanthaster) on coral reefs of Koh Tao, a diving destination in the Gulf of Thailand." Raffles Bulletin of Zoology 65 (2017).

Weterings, R. (2011). A GIS-based assessment of threats to the natural environment on Koh Tao, Thailand. Agriculture and Natural Resources, 45(4), 743-755.

www.ingramcontent.com/pod-product-compliance
Lightning Source LLC
Chambersburg PA
CBHW061113070526
44583CB00027B/3276